Understanding Randomness

LECTURE NOTES IN STATISTICS

A Series Edited By

D. B. OWEN

Department of Statistics
Southern Methodist University
Dallas, Texas

Additional Volumes in Preparation

Understanding Randomness

EXERCISES FOR STATISTICIANS

DAVID SALSBURG
Research Advisor
Pfizer Central Research
Groton, Connecticut

CRC Press
Taylor & Francis Group
Boca Raton London New York

CRC Press is an imprint of the
Taylor & Francis Group, an **informa** business

First published by Marcel Dekker, Inc.

Published 2020 by CRC Press
Taylor & Francis Group
6000 Broken Sound Parkway NW, Suite 300
Boca Raton, FL 33487-2742

First issued in paperback 2020

© Taylor & Francis Group, LLC
CRC Press is an imprint of Taylor & Francis Group, an Informa business

No claim to original U.S. Government works

ISBN 13: 978-0-367-58037-7 (pbk)
ISBN 13: 978-0-8247-7057-0 (hbk)

Visit the Taylor & Francis Web site at
http://www.taylorandfrancis.com

and the CRC Press Web site at
http://www.crcpress.com

Library of Congress Cataloging in Publication Data

Salsburg, David, [date]
   Understanding randomness.

   (Lecture notes in statistics ; v. 6)
   1. Random variables--Problems, exercises, etc.
2. Distribution (Probability theory)--Problems,
exercises, etc. 3. Multivariate analysis--Problems,
exercises, etc. I. Title. II. Series: Lecture notes
in statistics (Marcel Dekker, Inc.) ; v. 6.
QA274.S34 1983   519.5   83-18830
ISBN 0-8247-7057-9

It is part of probability that many
improbable things will happen.

Aristotle, POETICS

PREFACE

This book of exercises is designed to be used in addition to a standard text or as an extracurricular activity by the student who, I assume, has had at least one semester of statistics. If that semester has been in mathematical statistics, the additional level of sophistication would be useful; however, nothing in this book assumes a strong mathematical background. Its purpose is to provide the student with a "feel" for patterns in numbers that might occur fortuitously as a result of random noise, as differentiated from patterns that are generated by underlying structure.

Graduate programs in statistics tend to swing between two extremes: purely theoretical courses in which the student manipulates the mathematical structures of probability distributions, and the applied courses where the student is thrown on the firing line of "real data," where no one knows the true underlying structures. This book is designed as a middle ground. The student is given well-behaved clouds of uncertainty and allowed to see how the dice fall when the underlying probabilities can be calculated correctly.

I learned to look at such clouds of uncertainty by reading the papers of Karl Pearson and Sir Ronald Fisher, who both were clearly concerned about these problems. My wife, Fran, taught me even more by constantly probing my graduate studies with questions like, "What does it all mean?" or "How do you know you are right?" Finally, long hours of conversation with the late H. Fairfield Smith at the University of Connecticut showed me that a good statistician really does know what is random and what is not.

David Salsburg

v

CONTENTS

Understanding Randomness

INTRODUCTION

This book is dedicated to the art of statistical analysis (as opposed to the mathematics behind such analysis). John Tukey has proposed the following "equation" to describe the nature of statistical analysis:

DATA = FIT + RESIDUAL

That is, one attempts to examine a set of data to determine an estimate of some underlying model (the "Fit") that describes its central tendency, plus a cloud of random scatter (the "Residual") that obscures the nature of that model. The art of statistics consists of deciding what components of the data can be used to fit the model and what components can be considered random noise about the model. There is never a single correct version of this dichotomy. How the data are divided depends upon the use that is to be made of the analysis, such as the perceived need for parsimony in the model. However, above all else, the data are divided by the statistician's gut feeling for what is random noise and what is not.

The art of understanding randomness is one that has to be learned and practiced. When he invented (or did he discover?) the correlation coefficient, Karl Pearson kept track of the numbers of nesting storks he saw in his daily walks about London and calculated the correlation between that sequence of numbers and the corresponding counts of babies born on those days. This is not because

Pearson believed that storks brought babies or that he sought to prove the null hypothesis that they were independent. Rather, he was interested in seeing how the sample correlation coefficient behaved as a random variable when the true correlation was zero. R.A. Fisher kept detailed records of the growth of his children and used some of those numbers as examples in his texts. A well known modern statistician used to sit in a subway car and count the number of people with eyeglasses and without, comparing the frequencies of subgroups to those predicted by binomial assumptions. The art of statistics has to be practiced.

There is another art that needs continuous practice. This is the art of playing a musical instrument. Practice for a musician consists, in part, of going over patterns of notes that tend to occur often in the pieces that might be played. A proficiency in such patterns leaves the musician free to concentrate on the meaning and interpretation of a given piece because his or her fingers have learned to move by themselves through such patterns. To aid the musician in such practice, there are exercise books that display the various patterns in a single key and ask the student to transcribe these patterns into all other keys, or to invert them, or lay one against another. For the flutist, such are the exercises of Barrere, for the pianist the exercises of Hanon. In the beginning of each practice session, the good musician limbers up on segments of these exercises. Seldom does the musician play all of them in all their permutations and transmutations. However, they are there for constant work, to aid in the art of music.

This book is an attempt to create such a set of exercises for the statistician. Because many of the ideas presented here may be new to the student, there is some explanation of the purpose and structure of each set of exercises. Otherwise, this is a book that is meant to be "played" rather than studied. I assume that the student has had at least two semesters of a standard undergraduate course in statistics. If the course was one in mathematical statistics that required calculus, the student may have an advantage of sophistication. However, there is nothing in this book that requires a deeper prior knowledge than may be found in the more elementary course.

All that is needed from the student is an ability to calculate a few simple test statistics and a willingness to "play."

The chapters of this book present tables of numbers and make no attempt to graph them or provide other non-numeric displays. The student is encouraged to create graphical displays of this data and to examine the patterns in the data in many different ways, but such activity is up to the student. The physical act of copying numbers and plotting points provides the student with time and impetus to think about the nature of the randomness at hand. For the same reason, this book is not presented in terms of the manipulation of precoded data by computer. One must labor in the vineyard to produce good wine.

When the student has mastered most of the material in this book (one should not think of "finishing" a book like this), there will still be much more to be learned about the art of statistics. Multivariate data will have to be approached, and students will need experience with "real life" problems. However, this book should provide them with a good start.

All but the final chapter deal with data generated on a computer, homogeneous in structure with one or two aberrations clearly defined, like the exercises of Hanon. But, no one would give a piano concert playing exercises by Hanon. The regularity of pattern has no interest to a listener. Similarly, it is the irregularity of real data that requires the art of the statistician. Therefore, this book has a final chapter of real data. These are a set of seemingly simple problems, where the exact nature of the problem is not too well defined and is open to artistic interpretation by the data analyst. Algebra texts published in the late 19th century often had a final chapter of problems that were designed to tease the student. Some of them are obvious to a good student. Others require ingenuity and skill a little beyond the merely good student. And, some may never have been solved even by the authors. Chapter 6 of this book is an attempt to revive this delightful practice.

REFERENCE TEXTS

It is assumed that the student who will use this text will have had
at least two semesters of statistics. Most of the statistical techniques called for in this book should be recognized by such a student, or he or she will be able to find them in the textbook used in
that course. However, some of the techniques called for may not be
found in a specific textbook, or the student may wish a more complete discussion of the technique. To aid in the search for such
explanations, there is a section to each chapter. These briefly
describe whatever statistical techniques have been called for, but
were not used in a previous chapter. References are supplied to
one or more of the following eight standard textbooks or sets of
tables:

REFERENCES

1. Bliss, C., *Statistics in Biology,* McGraw-Hill Book Company, New
    York, 1970.

2. Dixon, W. and Massey, F., *Introduction to Statistical Analysis,*
    3rd Ed., McGraw-Hill Book Company, New York, 1969.

3. Hollander, M. and Wolfe, D., *Nonparametric Statistical Methods,*
    John Wiley and Sons, New York, 1973.

4. Kendall, M. and Stuart, A., *The Advanced Theory of Statistics,*
    Hafner Publishing Company, New York, 1967.

5. Miller, I. and Freund, J., *Probability and Statistics for
    Engineers,* 2nd Ed., Prentice-Hall, Englewood Cliffs, NJ, 1977.

6. Mood, A. and Graybill, F., *Introduction to the Theory of
    Statistics,* McGraw-Hill Book Company, New York, 1963.

7. Owen, D. B., *Handbook of Statistical Tables,* Addison-Wesley
    Publishing Company, Reading, MA, 1962.

8. Raktoe, B. and Hubert, J., *Basic Applied Statistics,* Marcel
    Dekker, Inc., New York, 1979.

1

UNIFORM DISTRIBUTIONS

INTRODUCTION

Statistical theory and practice are based upon the recognition that
"randomness" can mean many things and be described by a wide range
of probability distributions. Perhaps the simplest form of randomness is the uniform distribution. In this chapter, that means a set
of two-digit random variables where every integer from 0 to 99 has
an equal probability of occurring. This equal probability property
describes the marginal distribution of the numbers in Tables 1.1,
1.2, and 1.3. That is, if you were to pick a single number "at
random" from any of those tables, you have one chance in a hundred
that it will be a given value chosen beforehand. However, statistical practice deals with collections of numbers, and a second
property of "randomness" concerns how well consecutive numbers are
"connected," how well one can predict the next number from a previous sequence of numbers. In Table 1.1, where there is no such
"connection," all values are statistically independent. In Tables
1.2 and 1.3, each number is correlated with the number that occurs
before it in the same row. For Table 1.2, the correlation is mild,
for Table 1.3 strong. The exercises in this chapter are designed
to give the student a feel for both the uniform distribution and
statistical independence.

TABLE 1.1 Uniformly Distributed Independent R.V.

86	97	15	55	24	8	34	90	94	42	27	38	10	79	90	28	41
4	4	50	95	7	6	34	70	92	4	64	50	57	57	10	62	79
26	24	2	25	50	89	71	98	89	52	58	10	53	0	17	10	76
78	31	14	25	62	73	51	29	11	23	97	96	58	40	12	89	41
96	85	78	72	93	65	52	13	8	31	23	86	56	77	31	72	30
32	21	9	6	85	77	15	60	43	59	77	60	28	72	94	16	4
24	80	30	3	73	31	16	42	17	52	18	40	6	19	11	30	89
72	97	25	70	33	50	10	71	0	25	2	71	83	65	0	0	13
4	1	69	80	78	89	42	18	2	3	63	36	98	1	60	77	53
57	40	57	62	9	40	85	59	2	80	18	33	90	78	13	84	23
37	96	27	97	10	49	99	66	91	62	41	31	23	41	8	22	55
92	41	27	13	69	3	44	76	55	1	67	25	76	42	3	27	70
68	42	92	3	65	52	17	94	36	38	20	63	50	65	97	14	40
25	68	65	19	44	6	74	22	53	25	58	93	94	77	26	63	33
98	47	18	37	25	49	33	85	42	24	7	26	1	15	2	71	45
39	21	75	85	91	13	29	81	16	88	70	8	91	59	36	0	34
44	85	77	85	51	76	91	94	15	3	53	49	87	93	98	45	21
25	0	15	35	43	98	45	95	77	74	51	60	17	74	44	40	76
50	40	77	25	33	62	49	45	39	52	39	64	89	22	2	58	19
40	68	90	52	44	36	36	66	0	48	10	88	94	61	48	78	25
68	0	32	80	24	9	62	61	8	8	11	57	2	70	17	78	84
71	41	83	5	96	81	75	45	17	16	5	0	27	98	47	42	50
70	15	72	69	12	21	8	39	46	17	31	55	98	30	11	33	21
94	47	45	45	80	26	30	85	37	78	68	6	81	72	0	44	16
82	66	89	77	28	19	50	35	94	68	36	94	2	77	15	72	17
62	97	99	71	85	54	88	65	96	28	69	67	26	48	4	20	26
91	58	81	51	82	68	46	91	13	21	85	49	22	87	0	33	20
2	66	32	33	60	9	43	85	65	54	15	42	63	5	12	56	50

TABLE 1.2 Uniformly Distributed Mildly Dependent R.V.'s

```
45  26  26  37  21  44  42  42  60  72  44  37  44  32  59  60  66
33  55  55  65  80  46  54  34  63  39  34  48  28  63  71  54  57
59  63  80  87  51  51  46  23  57  65  63  77  53  39  56  64  47
63  50  74  45  29  45  25  34  32  23  58  37  23  47  47  27  16
43  53  55  77  48  33  37  47  25  17  23  31  55  75  56  40  39
52  65  77  86  72  56  34  62  36  54  76  42  28  26  59  76  41
56  33  55  40  24  23  38  42  44  34  44  35  19  20  43  55  61
37  31  29  52  61  59  47  48  34  19  11  38  33  62  36  22  48
29  63  45  51  70  65  69  68  72  85  65  34  27  23  61  43  50
44  37  21  40  68  63  53  68  42  67  59  55  64  36  55  76  51
84  51  57  45  45  26  18  12  54  74  63  43  62  65  74  64  73
60  40  51  32  32  28  23  36  28  47  34  18  11  45  31  40  62
28  58  63  35  33  51  47  44  27  14  13  10  53  53  44  38  50
51  40  48  29  63  51  57  73  38  31  62  78  60  47  32  53  63
27  29  27  41  59  76  48  54  61  36  49  67  38  60  44  43  38
46  33  40  45  29  32  40  62  46  25  19  23  25  25  45  70  54
41  44  66  50  31  33  22  55  42  24  25  48  40  69  39  27  30
57  77  63  35  62  75  70  62  38  23  29  50  51  66  41  40  52
23  59  44  26  19  49  59  68  67  70  44  69  76  78  76  38  34
49  73  42  62  65  75  83  76  70  80  82  61  57  35  66  52  62
55  74  63  58  33  18  10  54  41  67  56  76  71  72  42  49  67
29  64  43  70  67  72  39  40  54  37  68  80  62  38  35  63  62
77  51  68  52  45  44  68  58  66  56  55  28  28  57  43  32  45
36  33  46  47  71  54  46  68  72  47  48  67  38  60  68  50  39
46  56  49  60  39  40  52  51  63  69  44  48  57  53  59  30  23
46  66  61  60  32  23  14  20  41  63  57  34  44  63  65  37  26
55  72  52  60  76  62  41  62  61  77  75  87  75  87  67  74  37
50  57  32  30  18  33  29  45  41  68  73  45  36  47  67  77  81
```

TABLE 1.3 Uniformly Distributed Strongly Dependent R.V.

8	23	27	42	43	45	42	51	55	53	51	53	49	50	56	64	70
71	70	69	64	68	57	58	53	49	45	44	49	45	45	49	54	62
59	48	48	52	57	50	47	43	41	34	29	25	26	23	37	34	41
46	51	59	62	67	58	49	48	55	55	54	59	59	52	49	50	43
44	48	47	41	39	44	44	54	58	63	60	67	67	59	48	50	53
45	43	54	51	54	53	53	48	47	51	49	40	40	46	39	38	31
50	47	42	36	36	40	48	58	47	54	46	44	37	34	36	47	42
46	50	52	50	50	42	39	36	46	40	40	44	47	53	62	67	70
71	64	52	42	53	60	51	47	46	50	41	53	54	59	62	56	49
64	67	59	57	57	60	53	59	51	50	45	48	57	46	45	51	55
58	65	66	63	64	70	68	65	64	68	64	56	53	45	54	49	45
52	47	47	48	57	63	60	61	66	61	63	60	56	52	42	35	29
36	38	35	31	39	47	47	49	51	50	53	44	43	50	41	47	56
53	52	60	58	51	49	50	58	56	46	50	45	46	55	49	56	58
46	47	38	32	33	41	35	46	53	44	37	41	36	29	38	31	33
35	42	37	36	31	43	39	37	31	41	44	51	50	55	58	54	51
55	54	50	56	55	56	56	48	56	62	52	58	54	54	52	45	50
42	48	54	46	42	40	33	39	36	32	43	41	44	49	40	48	46
41	52	49	58	46	38	48	44	52	54	48	44	36	35	35	33	29
44	44	37	45	53	62	68	72	62	52	52	48	43	48	44	54	60
52	44	41	41	39	34	31	35	35	30	31	43	34	45	53	48	56
39	33	35	42	48	56	64	59	61	56	53	53	45	54	46	44	51
48	55	46	54	55	50	57	63	70	66	65	72	62	69	61	68	60
59	55	57	47	58	54	50	57	48	55	47	56	62	52	50	60	59
50	57	52	57	48	51	46	46	42	50	55	46	48	50	54	55	55
48	42	36	33	35	48	55	48	48	48	46	38	43	49	53	53	54
62	55	59	56	58	57	55	57	58	65	56	51	53	48	54	45	52
44	41	48	50	52	49	43	48	55	48	48	51	53	58	51	60	48

EXERCISES

Exercise 1.1:   Browsing

Table 1.1 consists of uniformly distributed independent two-digit
numbers. Examine it for "patterns." The human eye is capable of
seeing patterns in a way that cannot be duplicated by any computer
program. Look for unusually high frequencies of specific digits or
pairs of digits. Look for unexpectedly frequent reoccurrences of
pairs of numbers. For instance, note how often a number ending in
"8" occurs, how often it seems to occur in a small block of consecutive numbers. Note how often numbers "close together" in value
seem too close together on the table.

Do not be influenced by the knowledge that these are independent uniform random numbers. There are patterns here, and a good

statistician should have a feel for what kinds and frequencies of patterns occur as the purely fortuitous result of random noise.

Exercise 1.2: Testing Frequencies--Null Hypothesis True

Pick a two-digit number at random from Table 1.1. Use the first digit to locate the row and the second one to locate the column at which to start this exercise in Table 1.1. Going from left to right and then down to the next line, count out 30 groups of five two-digit numbers, broken into subsets of 10 groups each. "Wrap around" to the beginning of the table if necessary. For each group of five numbers, tally how many of them are even, so that each subset of ten groups will give you a certain number of groups with 0, 1, 2, 3, 4, or 5 even numbers. You will then have samples of size 10, 20 (two subsets), and 30 (three subsets) of a binomial variable with $n = 5$ and $p = 0.5$. Use two different tests of goodness of fit (such as the Pearson Chi Square, Likelihood Ratio, Discrete Kolomogorov) to test the hypotheses:

1. Binomial with $n = 5$ and $p = 0.5$.
2. Binomial with $n = 5$ and p estimated from the data.
3. Normally distributed with mean 2.5 and variance 1.25.

EXERCISE 1.3: Testing Frequencies--Null Hypothesis Moderately False

As in Exercise 1.2, pick a two-digit number at random from Table 1.1 to identify the starting place in Table 1.1 and count out 30 groups of five numbers each ("wrapping around" to the beginning of the table if necessary). Tally the number of times each group of five has 0, 1, 2, 3, 4, or 5 numbers less than 31. As in Exercise 1.2, collect these tallies into subsets so that you have one subset of 10, two subsets (20), and all three subsets (30). Test the following hypotheses with at least one goodness of fit test:

1. Binomial with n = 5 and p = 0.5.
2. Binomial with n = 5 and p estimated from the data.
3. Normally distributed with mean 2.5 and variance 1.25.
4. Normally distributed with mean 2.5 and variance estimated from the data.

Exercise 1.4: Testing Frequencies--Null Hypothesis Markedly False

As in Exercises 1.2 and 1.3, choose a number at random from Table 1.1 and let it identify your starting place. Count out 30 groups of five numbers in Table 1.1 ("wrapping around" to the beginning if necessary). Tally the numbers of groups with 0, 1, 2, 3, 4, or 5 numbers less than 11. As in Exercises 1.2 and 1.3, collect these tallies into subsets of 10. Then, apply goodness of fit tests to samples of size 10, 20 (two subsets), and 30 (all three subsets) to test the hypotheses:

1. Binomial with n = 5 and p = 0.5.
2. Binomial with n = 5 and p estimated from the data.
3. Normally distributed with mean 2.5 and variance 1.25.
4. Normally distributed with mean 2.5 and variance estimated from the data.
5. Poisson with mean 0.5.

Exercise 1.5: Looking at Runs

This exercise should be applied to Tables 1.1 (purely random), 1.2 (mildly correlated), and 1.3 (strongly correlated). For each table, pick a number at random from Table 1.1 and let the first digit define the starting row and the second digit the starting column. Examine the sequence of 30 two-digit numbers that follow that point (going from left to right then down to the next row and "wrapping around" to the beginning of the table if necessary). For each of the three sets of 30 numbers (one from each table), pick 50.5 as a median break-point and tally (a) the number of runs above and below 50.5, (b) the number of items ($N(1)$) above and the number of items

($N(2)$) below 50.5, and (c) the length of the longest run (either above or below 50.5).

Run the following hypothesis tests:

1. Using $N(1)$ and $N(2)$ only construct a normal theory test that Prob(X less than 50.5) = 1/2.
2. Using the number of runs and $N(1)$ and $N(2)$, test the hypothesis that the sequence consists of independently distributed random variables with median 50.5.
3. Using the length of the largest run and $N(1)$ and $N(2)$, test the hypothesis as in 2. above.

Exercise 1.6: More Browsing

Tables 1.1, 1.2, and 1.3 differ in their autocorrelation from one number to the next. They were constructed in the following fashion: Running from left to right on a single row, each number is of the form

$$X(i) = rU + (1-r)X(i-1)$$

where U is a uniform random digit between 0 and 100 and (1-r) is set at 0.5 for Table 1.2, 0.85 for Table 1.3, and 0.0 for Table 1.1. Going down on a single column means that the value below is related to the one above with a similar form but with the value of "1-r" the 17th power of the original coefficient (and the value of "r" set accordingly). Thus, for Table 1.2, successive values going down a column are almost completely independent, while for Table 1.3 successive values going down a column have a level of dependence that is barely detectable. Keep this structure in mind in what follows.

Start by comparing Tables 1.1 and 1.3. If you have done Exercises 1.1 and 1.5, you will have some initial feel for the patterns of pure randomness (Exercise 1.1) and how correlation affects the patterns of runs (Exercise 1.5). As you compare Tables 1.1 and 1.3, look for dissimilarities. Most of these will be due to the tendency for successive numbers in Table 1.3 to be close to each other in value. Formulate some simple measures of what you see in those

comparisons and compute them for parts of Table 1.2. One such measure is the longest run of values within X units of each other for
various choices of X. The idea here is to see how you might reduce
complicated patterns of comparison to single numbers that might con-
tain much of the information in those comparisons. The measures can
be developed on Table 1.3, tried out on Table 1.1 to see how easily
they can be fooled, and, finally, applied as test measures on Table
1.2, to get some idea of their power to discern serial correlation.

If you have the software available, you will want to run some
standard tests of serial correlation on runs of differing lengths
from Tables 1.1, 1.2, and 1.3.

SELECTED WORKED-OUT EXAMPLES OF EXERCISES

The following pages are worked out examples for Exercises 1.2 and
1.5. Exercise 1.1 has no formal structure to it, so there is no way
of showing how it "should go." Exercise 1.2 is worked out completely, to aid the student who may never have worked with goodness
of fit tests before. Exercise 1.5 is worked out in part (for a
single table), and is typical of the way in which it would be worked
on other tables. Finally, Exercise 1.6 is, like Exercise 1.1, a
browsing exercise that has no correct "solution."

Worked-out Example of Exercise 1.2

Groups of two-digit random numbers chosen from Table 1.1

Group	Numbers				
1	19	29	49	73	54
2	82	2	28	85	4
3	60	92	67	71	2
4	24	39	39	59	29
5	6	17	65	37	95
6	81	32	37	69	90
7	8	93	4	39	93
8	81	71	0	78	41
9	82	37	91	64	2
10	81	3	79	12	13

Group	Numbers				
11	79	39	7	95	65
12	27	18	58	58	69
13	42	44	5	43	40
14	76	46	42	77	5
15	71	4	4	34	28
16	68	0	13	18	61
17	29	78	54	45	14
18	71	14	10	63	88
19	36	54	65	26	26
20	99	38	93	6	85
21	38	56	75	26	25
22	20	51	2	98	73
23	70	11	31	93	16
24	17	24	7	52	88
25	90	28	45	58	79
26	2	49	35	25	36
27	42	36	39	55	70
28	0	28	60	41	6
29	62	3	81	42	54
30	81	46	27	55	56

Test that the number of even values is binomially distributed with $n = 5$ and $p = 0.5$:

j	$p(j)$	$n = 10$ E(f)	f	$n = 20$ E(f)	f	$n = 30$ E(f)	f
0	0.0312	0.312	0	0.624	1	0.936	1
1	0.1562	1.562	4	3.124	4	4.686	4
2	0.3125	3.125	3	6.250	4	9.375	6
3	0.3125	3.125	2	6.250	8	9.375	15
4	0.1562	1.562	1	3.124	3	4.686	4
5	0.0312	0.312	0	0.624	0	0.936	0
Chi-Square 5 d. of f.		5.04		2.40		5.73	

Test that the counts of even values is binomially distributed with $n = 5$ and p estimated from the data:

$n = 10$ p estimated at 0.400

j	$p(j)$	$E(f)$	f
0	0.0778	0.778	0
1	0.2592	2.592	4
2	0.3456	3.456	3
3	0.2304	2.304	2
4	0.0768	0.728	1
5	0.0102	0.102	0

Chi square with
5-1=4 d of f.
1.82[a]

$n = 20$ p estimated at 0.480

j	$p(j)$	$E(f)$	f
0	0.0380	0.760	1
1	0.1755	3.510	4
2	0.3240	6.479	4
3	0.2990	5.981	8
4	0.1380	2.760	3
5	0.0255	0.510	0

Chi square with
4 d of f.
2.31[a]

$n = 30$ p estimated at 0.513

j	$p(j)$	$E(f)$	f
0	0.0274	0.822	1
1	0.1443	4.328	4
2	0.3040	9.119	6
3	0.3202	9.606	15[b]
4	0.1686	5.059	4
5	0.0355	1.066	0

Chi square with
4   of .
5.45[a]

[a]Note how estimate of p reduced the absolute value of
the chi-square statistic, but that the number of
degrees of freedom is reduced to compensate for this.

[b]Note how the "imbalance" in the set of 30 still exists
(9.6 predicted versus 15 observed) even when the underlying value of p is estimated from the data.

Test that the counts of even values is normally distributed
with mean 2.5 and variance 1.25, where $Prob(X=j) = F(z(j+ 1/2)) - F(z(j- 1/2))$, and $z(t) = (t-2.5)/sqr(1.25)$.

j	$z(j+1/2)$	$p(j)$	$n = 10$ $E(f)$	f	$n = 20$ $E(f)$	f	$n = 30$ $E(f)$	f
0	-1.789	0.0368	0.368	0	0.736	1	1.104	1
1	-0.894	0.1489	1.489	4	2.978	4	4.467	4
2	0.000	0.3143	3.143	3	6.286	4	9.429	6
3	0.894	0.3143	3.143	2	6.286	8	9.429	15
4	1.789	0.1489	1.489	1	2.978	3	4.467	4
5	infin.	0.0368	0.368	0	0.736	0	1.104	0
Chi-square			5.55		2.48		5.75	

Note how the chi-square goodness of fit statistics tend to be closer for the fit to a normal and binomial the larger the sample size.

Exercise 1.5 Applied to Table 1.2

Random number off Table 1.1 = 82: Start in Table 1.2 at the eighth row second column, resulting in the following set of numbers:

31 29 52 61 59 <u>47 48 34 19 11 38 33</u> 62 <u>36 33 48 29</u>

63 <u>45</u> 51 70 65 69 72 85 65 <u>34 27 23</u>

$N(1)$ = number of values less than 50.5 = 17
$N(2)$ = number of values greater than 50.5 = 13
number of runs of values below 50.5 = 5
number of runs of values above 50.5 = 4
total number of runs = 9,   length of largest run = 8

Test of whether $N(1)$ is distributed as a binomial with $n = 30$ and $p = 0.5$.

Normal approximation: mean = 30(0.5) = 15
variance = 30(0.5)(0.5) = 7.5
$z = (N(1)-15))/\text{sqr}(7.5) = 0.730$

If R = number of runs, then R is approximately normal with

$$\text{mean} = (2N(1)N(2))/(N(1)+N(2)) + 1 = 15.73$$
$$\text{variance} = (2N(1)N(2)(2(N(1)N(2)-N(1)-N(2)))) /$$
$$((N(1)+N(2))^2(N(1)+N(2)-1)) = 6.977$$
$$z = (R-15.73)/\text{sqr}(6.977) = -2.549$$

Thus, we cannot reject the hypothesis that the numbers in Table 2.1 have a marginal distribution with values less than 50.5 having a probability of 0.5, but we can clearly reject the hypothesis that they are statistically independent since there are too few runs of values above and below 50.5 in this set of 30 values.

Length of longest run = L = 8: If N = 30 and if the observations are independent, then

Prob(L is greater than or equal to 5) = 0.50
Prob(L is greater than or equal to 7) = 0.08
Prob(L is greater than or equal to 9) = 0.02

Thus, L = 8 implies significance near 0.05. For a complete table of the distribution of the longest run, see Olmstead, *Ann. Math Stat. 10,* 29-35 (1958).

STATISTICAL TECHNIQUES AND REFERENCES
Pearson Chi-Square Goodness of Fit Test

Data are divided in k cells with $f(i)$ observations in the ith cell, i = 1,2,...,k. Under the null hypothesis, the expected number of observations in the ith cell is $e(i)$.

The sum of the elements

$$\Sigma(f(i)-e(i))^2/e(i)$$

has an asymptotic chi square distribution with degrees of freedom d, where d lies between $(k-1)$ and $(k-1-p)$, and p is the number of parameters estimated from the data.

References

Dixon and Massey, Sections 13-4, 13-5, pp. 243-249.
Kendall and Stuart, sections 30.8-30.23, pp. 423-433.
Mood and Graybill, Section 12.10, pp. 308-311.
Miller and Freund, Chapter 9, pp. 256-258.
Raktoe and Hubert, Section 13.4, pp. 245-250.

Likelihood Ratio Test

The null hypothesis $H(0)$ is embedded in a more general hypothesis H. The density function has the same form for all elements of the general hypothesis and differs only with respect to the values of some set of parameters $\underline{s}$.

$$\text{density} = f(x;\underline{s})$$

The likelihood for the set of observed data is

$$L = \text{the product of elements } f(x(i);\underline{s})$$
$$L^* = \log(L)$$

The likelihood ratio test statistic is

$$-2\ (\max(\text{over } H(0))\ L^* - \max(\text{over } H)\ L^*)$$

It is asymptotically distributed as a chi square with degrees of freedom equal to $(p-k)$ where

p = dimension of the free parameter in H, and
k = dimension of the part of the parameter space fixed by $H(0)$

References

Kendall and Stuart, Chapter 24.
Mood and Graybill, Section 12.6, pp. 296-301.

KOLOMOGOROV-SMIRINOV TESTS

Under the null hypothesis, let

$F(t) = \text{Prob}(X \text{ less than } t),$

and let the empirical distribution function be

$Sn(t) = \text{proportion of observations less than or equal to } t.$

Then, the test statistic is

$d = \text{Max} \mid F(X(i)) - Sn(X(i)) \mid \text{ taken over } X(1), X(2), \ldots, X(n).$

References

Hollander and Wolfe, Chapter 10, pp. 219-229.
Kendall and Stuart, Sections 30.49-30.62, pp. 452-460.
Miller and Freund, section 10.5, pp. 285-287.
Owen, Section 15, pp. 423-442.

When testing for normality, but without knowing the true mean and variance, the Kolomogorov Test has been modified by

Lilliefors, H., On the Kolomogorov-Smirinov Test for Normality with Mean and Variance Unknown, *J. Am. Stat. Assoc.*, *62*, 399-402, 1967.

RUNS TESTS ABOVE AND BELOW SOME FIXED VALUE
(SUCH AS THE MEDIAN)

Observed data are ordered. A sequence of consecutive values above (or below) the fixed value is run. If

$N(1) = \text{the number of observations below the fixed value}$
$N(2) = \text{the number of obs. above the fixed value}$

then the number of runs is asymptotically normal with

$$\text{mean} = 2N(1)N(2) \;/\; (N(1)+N(2)) \;+\; 1, \text{ and}$$

$$\text{variance} = (2N(1)N(2)(2N(1)N(2)-N(1)-N(2)))/$$
$$((N(1)+N(2))^2 \; (N(1)+N(2)+1))$$

If $N(1) = N(2)$ and is large, then the average length of a run lies between 3 and 5 and the maximum length of a run tends to be between 10 and 13.

References

Dixon and Massey, Section 17-3, pp. 342-344.
Kendall and Stuart, Exercise 30.8, p. 463.
Mood and Graybill, Section 16.4, pp. 409-416.
Miller and Freund, Section 10.4, pp. 282-285.
Owen, Sections 12.4, 12.5, pp. 373-382.

2

SINGLE SAMPLES OF CONTINUOUSLY DISTRIBUTED RANDOM VARIABLES WITH POSITIVE SKEW

INTRODUCTION

A frequent problem with experimental data deals with "outliers." Most good experimentalists have some idea of the expected outcome of their experiments. Unexpected outcomes are looked upon with suspicion. When a sequence of experiments have been conducted, the experimentalist is often concerned that one or more of the results may be abherrant. Thus, the search of the data for "outliers," numbers that are not typical and which will tend to skew the conclusions away from the true central tendency. This chapter deals with data that may or may not have such outliers lying on one side of the true central tendency. There are four sets of data in Tables 2.1, 2.2, 2.3, and 2.4. Table 2.1 is symmetric about its mean. Tables 2.2, 2.3, and 2.4 have increasing degrees of assymmetry. All four tables have the same mean (50) and the same variance (100). Only the skewness differs.

EXERCISES
Exercise 2.1: Browsing

Look through Table 2.1 for numbers greater than 65 (1.5 standard deviations above the mean). Look for numbers less than 35. Note

TABLE 2.1 Symmetric R.V.'s, Mean=50, Variance =100

52.0	49.1	30.0	41.4	49.1	42.8	39.2	35.9	72.8	53.1
40.5	69.1	39.8	52.4	48.9	45.7	33.6	63.7	47.0	39.7
51.2	37.6	50.4	51.2	65.8	38.2	59.2	49.9	47.7	44.7
56.7	35.4	51.0	39.9	54.9	54.1	47.6	46.5	46.9	54.0
37.8	58.9	36.2	35.0	58.4	63.3	52.8	37.0	47.0	33.5
45.9	74.8	53.5	44.2	44.4	39.4	38.4	61.1	53.1	26.7
48.0	37.8	46.5	72.5	46.5	49.5	29.4	55.4	45.0	43.7
54.2	45.1	66.9	37.8	45.2	48.8	71.3	43.4	69.7	44.1
62.0	47.5	35.8	68.5	24.1	51.2	57.7	46.2	57.4	43.9
57.2	51.4	46.5	45.7	42.2	38.8	57.9	46.5	42.5	60.5
54.0	60.8	51.9	70.6	53.6	31.0	47.7	45.9	50.2	46.4
56.7	63.6	46.0	59.7	48.2	56.3	43.5	50.0	53.8	43.5
14.0	52.9	44.4	49.8	62.6	37.5	33.7	40.9	40.2	59.9
48.4	65.0	52.7	54.3	56.1	52.6	26.7	40.1	49.3	52.1
42.1	56.7	46.5	52.6	41.6	64.3	60.1	44.1	67.5	42.5
51.7	33.3	54.6	45.2	43.3	47.8	50.8	72.5	36.3	36.2
43.6	49.3	57.0	55.1	64.7	41.3	18.7	43.4	60.4	27.4
61.3	63.5	49.2	68.9	37.1	68.7	36.8	56.0	54.2	48.1
37.3	61.0	55.8	48.4	61.2	38.1	57.6	51.9	61.3	50.7
50.9	52.5	64.1	54.7	63.5	67.3	49.0	39.5	56.8	46.1
54.3	45.4	50.2	57.0	49.6	38.2	42.2	49.9	55.4	50.1
68.3	45.3	69.7	38.2	36.1	51.9	61.3	49.5	36.7	59.1
49.8	42.5	43.7	68.1	62.9	56.5	48.6	38.8	50.1	46.7
43.1	33.8	53.8	57.3	58.2	59.7	46.5	60.3	50.7	54.6
26.2	42.4	56.0	53.0	27.0	47.6	57.9	37.4	47.8	47.7

TABLE 2.2 Mildly Assymmetric R.V.'s, Mean = 50, Variance = 100

72.7	41.4	32.6	53.1	49.9	43.1	55.9	46.4	37.7	54.4
48.7	56.2	56.2	37.8	60.4	46.5	45.0	48.8	31.2	42.5
57.3	34.5	41.1	67.8	34.7	65.0	46.3	55.3	34.6	38.9
44.2	42.2	55.1	67.4	63.6	40.9	45.6	60.1	56.5	58.1
31.3	41.2	49.0	44.3	66.1	46.0	46.8	68.8	43.6	45.8
56.9	56.5	41.7	58.9	53.8	59.9	35.2	51.5	51.1	59.6
55.1	48.6	47.4	50.8	40.5	69.3	50.0	54.7	38.8	42.5
33.5	54.5	47.8	47.6	59.4	54.8	56.2	78.3	34.0	53.4
62.3	55.0	43.7	45.2	63.2	49.2	38.1	63.3	38.3	48.9
72.8	54.4	52.4	38.3	50.2	50.8	57.3	37.8	45.0	59.2
42.9	40.1	46.4	44.2	39.4	42.4	63.9	50.5	55.5	37.5
64.1	40.6	51.4	66.9	45.7	68.2	59.4	49.0	54.8	40.0
62.2	53.1	69.8	45.1	38.6	49.4	45.8	60.5	55.7	60.2
37.9	45.8	62.1	63.9	48.7	48.7	61.6	37.0	43.8	46.2
56.5	62.8	46.4	52.0	44.0	54.4	56.5	55.6	56.6	52.5
43.0	37.2	40.9	40.6	49.1	54.8	46.5	54.4	63.7	61.3
39.1	42.2	48.2	49.1	28.8	37.0	60.2	34.8	60.4	60.9
39.8	58.1	44.0	43.5	48.4	64.3	50.2	52.3	33.8	35.0
47.0	53.3	54.7	59.5	56.5	56.8	45.5	54.3	47.0	41.5
53.8	57.3	62.4	59.2	33.1	49.8	79.1	59.9	44.6	37.0
56.8	47.0	56.0	56.8	62.0	56.4	54.8	40.2	53.5	61.7
53.2	42.0	72.9	52.4	37.7	24.5	25.5	55.5	56.9	43.3
38.2	68.4	44.3	53.8	69.8	55.0	48.2	56.1	52.8	39.9
48.9	32.5	50.0	40.1	36.6	61.4	49.8	57.0	25.7	60.2
41.0	27.0	58.8	57.6	49.0	34.9	46.0	36.6	54.7	39.3

TABLE 2.3 Moderate Assymmetry, Mean=50, Variance=100

60.7	75.1	62.9	58.1	66.3	42.7	51.1	36.4	105.0	66.8
48.7	53.5	80.7	51.1	63.6	64.1	46.7	17.0	42.0	57.5
43.6	57.1	32.8	67.9	29.5	40.8	52.1	82.5	34.4	19.1
63.2	27.0	70.1	70.5	43.9	94.7	44.3	82.1	40.4	35.6
45.7	76.5	34.7	33.8	41.1	58.4	21.1	85.5	48.9	55.2
22.8	101.6	39.2	66.5	68.1	88.6	66.8	70.5	49.5	70.3
49.2	43.3	82.7	46.4	50.5	67.3	35.0	61.6	21.6	24.0
50.0	87.8	45.3	66.9	61.5	62.1	38.5	115.0	78.7	25.4
12.6	52.4	57.0	15.7	88.4	48.2	41.2	46.9	49.9	58.4
34.3	23.1	43.9	63.6	27.6	57.6	62.3	27.9	63.0	39.0
43.0	76.3	54.2	40.9	43.2	23.0	0.9	57.4	54.8	77.1
48.4	32.5	38.1	62.0	39.1	31.7	48.7	43.0	100.9	87.0
58.1	95.1	34.7	37.2	98.8	70.4	43.1	64.4	84.0	68.9
68.5	50.1	79.1	74.2	73.0	51.7	15.0	47.9	14.5	48.2
38.1	83.7	60.1	59.7	52.1	80.2	41.9	55.5	73.1	40.8
59.4	46.3	29.2	43.1	32.0	58.6	40.8	56.9	39.7	62.5
61.7	57.8	55.9	35.0	96.3	49.3	64.1	71.7	38.1	43.8
61.1	68.8	47.8	60.6	54.1	36.7	44.8	52.9	37.4	58.7
48.2	77.9	64.5	27.2	99.9	69.7	62.3	97.8	60.8	56.7
55.3	14.1	56.2	20.9	32.9	47.6	54.4	52.5	65.3	2.3
29.5	50.7	69.6	46.6	39.6	72.5	35.0	64.0	44.4	28.2
51.6	57.1	80.8	50.0	86.2	39.2	57.5	48.5	32.0	85.5
51.2	20.2	49.7	49.4	61.3	30.1	44.4	92.0	43.1	47.1
77.8	46.0	34.1	53.5	66.7	47.4	52.8	46.9	46.3	12.2
84.7	25.8	78.6	25.5	34.3	44.0	38.7	34.9	35.4	36.3

TABLE 2.4 Extreme Assymmetry, Mean=50, Variance=100

43.4	36.8	52.0	17.3	35.7	128.4	84.0	18.2	-2.9	27.8
129.3	77.5	58.2	123.8	63.3	133.0	75.4	39.2	46.0	52.7
60.6	2.2	14.7	62.9	-11.3	60.4	47.5	46.5	33.2	48.8
30.2	88.6	49.6	45.1	37.1	49.5	82.4	82.4	31.3	43.0
13.6	74.6	12.2	52.2	22.0	96.4	37.7	47.9	27.8	60.7
-8.9	55.5	99.3	38.6	1.7	40.2	56.9	-16.3	56.5	70.3
41.2	49.9	19.3	47.2	18.6	-3.0	40.6	121.8	4.0	54.1
50.2	24.6	81.0	142.2	85.1	17.5	-14.8	77.1	5.6	9.7
89.3	99.8	45.6	41.7	96.6	-13.3	53.7	-25.9	-1.4	47.5
45.7	96.5	45.3	97.3	47.3	-10.3	39.2	57.8	18.5	23.0
65.7	55.6	24.4	3.9	83.1	27.1	33.3	41.9	82.8	108.6
1.1	58.6	40.0	81.4	21.3	62.7	41.8	104.2	55.3	87.2
162.8	20.6	11.4	23.8	80.6	88.2	47.0	5.1	34.1	87.2
61.5	27.1	32.7	39.8	78.3	33.2	74.9	63.6	6.1	36.5
23.3	98.0	49.2	29.1	8.3	21.1	29.3	41.0	57.3	54.2
57.1	124.0	10.6	59.5	59.8	84.2	62.9	42.0	86.8	31.0
48.0	38.1	121.4	22.3	78.7	15.8	59.7	69.0	26.4	54.8
55.2	14.0	40.4	66.6	39.3	20.7	55.4	40.8	33.7	20.5
10.9	10.1	59.7	56.9	26.3	76.6	40.0	49.6	90.1	-12.6
20.7	117.2	31.3	25.6	61.1	56.4	44.6	71.0	78.6	26.0
-21.4	66.8	124.3	60.6	83.9	112.8	52.9	0.0	57.8	48.6
23.9	100.5	94.9	31.9	51.3	58.5	49.1	45.6	-6.0	19.1
100.8	28.3	53.9	28.8	31.7	59.6	101.3	25.4	39.7	-2.2
136.9	17.1	49.3	46.3	67.5	22.2	100.2	139.8	108.7	16.6
71.2	7.8	37.8	59.5	65.1	8.2	66.3	62.3	7.2	61.5

how or whether the two "tails" tend to be the same size. Look for
more extreme outliers (greater than 70, less than 30). Look for
patterns among their occurrence, gaps between them, runs of outliers,
anything else your eye might see as "unusual."

Having established some "feel" for the occurrence of outliers
in a well behaved sample, turn to Table 2.4. Note how the extreme
skew of this table leads to a bunching up of numbers below the mean
(some even going negative) with the outliers stretched out and having large gaps between them. What other patterns do you see? Are
there any patterns in their occurrence on the page, runs of outliers,
large areas without outliers? Remember that the data in this table
are realizations of independent identically distributed random
variables. Any spatial patterns your eye sees on the page are
purely random noise.

Finally, using the "feel" for symmetry and assymmetry gained in
Tables 2.1 and 2.4, examine Tables 2.3 and 2.2 for outliers, bunching up of values below the mean, and other patterns your eye has
seen.

Exercise 2.2: More Browsing

Start at random in Table 2.1. Copy out sequences of numbers of
length 10. Note how many of these sequences have individual
"outliers" (numbers greater than 65). Note how many have apparent
imbalances between values above and below 50 (the mean). Do the
same for Tables 2.2, 2.3, and 2.4. Can you use the insights gained
this way to distinguish data from Tables 2.1 and 2.4, between 2.1
and 2.3, between 2.1 and 2.2?

Exercise 2.3: Testing for the Mean

This exercise should be run on each of the four tables in this
chapter.

Starting at random, draw a sample of size 50 ("wrapping around"
to the beginning of the table if necessary). Work with the first 10,
then with the first 30, then with all 50. Compute the mean, median,

and at least one trimmed mean. Compare these results as a function
of the table used and note their increasing stability as you go from
sample sizes of 10 to 30 to 50. (Remember, the true underlying mean
is 50.) Using both Wilcoxon Signed Rank and t-tests, test the
hypotheses:

1. Central Tendency = 50;
2. Central Tendency = 60;
3. Central Tendency = 70.

As you run through this exercise, be aware of the scatter of
individual numbers. Before you compute the final test statistic,
try to guess at the significance level that will emerge for each of
the three hypotheses. Note the similarities and dissimilarities of
results between the nonparametric Wilcoxon test and the normal
theory t-test.

Exercise 2.4: Estimating Variance

As you did Exercise 2.3, you were probably aware of the effect of
individual outliers on the denominator of your t-test. This exercise explores that phenomenon more closely.

Starting at random in each table, pick samples of size 50
("wrapping around" to the beginning of the data if necessary). For
each of the following estimators, compute the result for the first
10, the first 30, and then all 50 numbers. You are looking, in this
exercise, for the effect of increasing sample size and varying
degrees of assymmetry on the stability of estimators of variance.

Compute each of the following estimators of variance:

1. The standard sample Variance, s^2;
2. The square of the BLUE normal theory estimator of the
   standard deviation;
3. An estimator based upon the range;
4. A Winsorized variance estimator.

Compare the resulting estimates.

Using the computed values of s^2 and the chi square distribution, test the hypothesis that the underlying variance is 100 and construct 90% confidence bounds on the underlying variance.

Exercise 2.5:   Transformation

The Wilcoxon test in Exercise 2.3 involved the use of a rank transformation. That is, all the original values were replaced by their ranks within the sample. There are other transformations designed to reduce the effects of skewness on data. The most widely used are the power transformations, and the logarithm (which is a limiting case). Run Exercise 2.3 over again, applying the same statistical tests to the random samples you pick after

1.   Converting all values to the square root;
2.   Converting all values to the natural logarithms;
3.   Converting all values to the inverse hyperbolic sine.

(The last function has the advantage that it can be applied to negative data, which cannot be done for (1) and (2).)

Compute confidence bounds on the central tendencies based on t-tests with these transformations. Compare these to confidence bounds based upon t-tests run on the original data. Note that the null hypothesis

$H(0)$:   Median(X) = 50

converts, when using a transformation $Y = g(X)$, to

$H(0)$:   Median(Y) = $g(50)$

Exercise 2.6:   Two Sample Comparisons

Starting at random in Table 2.1, draw two samples of size 50 ("wrapping around" to the beginning if necessary). Call the first sample X and the second sample Y. Run the following procedures for the first 10 values of X and for the first 10 values of Y, then for the first 30 values of X and the first 30 values of Y, and finally for all 50 values of X and all 50 values of Y.

Order the combined sample of X and Y, keeping X in one row and Y in another row to aid in comparison. Note the tails of the ordering, that is, how many values of X or Y stand alone in the extreme ends of the ordering.

Run Wilcoxon rank sum tests and two sample t-tests comparing X and Y. Compute 90% confidence bounds on the differences in central tendency between X and Y, based upon the Wilcoxon tests and then upon the t-tests. Is the extra effort involved in the Wilcoxon confidence bounds worth it in terms of increasing the tightness of the bounds?

Repeat this exercise for Tables 2.2, 2.3, and 2.4.

STATISTICAL TECHNIQUES AND REFERENCES
Best Linear Unbiased Estimate (BLUE) of Standard Deviation

This is the linear function of the ordered data that has the minimum variance among all unbiased linear functions of ordered data. It assumes that we know the underlying family of distributions, since the coefficients of the minimum variance unbiased estimate vary from one family to another. If the underlying distribution is normal, the coefficients are found in the references.

References

Dixon and Massey, Section 9-5, pp. 138-139.
Kendall and Stuart, Sections 19.6, 19.7, pp. 79-81.
Owen, Sections 6.1-6.1, pp. 138-142.

RANGE ESTIMATES OF STANDARD DEVIATION

These are "rough and ready" estimates of dispersion of the form

$$K \ (\max X(i) - \min X(i) \ )$$

Depending upon how much the analyst wishes to assume about the underlying distribution, different choices of K are available. $K =$ 1/4 is often used on the assumption that most of the data in a small set will fall within two standard deviations of the mean.

References

Dixon and Massey, Section 9-5, pp. 136-138.

WINSORIZED ESTIMATE OF VARIANCE

The set of data is ordered and the k largest values are replaced by the value that is kth from the end, while the k smallest values are replaced by the kth from the beginning. The standard sample variance S^2 is them computed from these "Winsorized" values. The new estimate of variance is created by multiplying the sample variance by the square of

$$(n-1)/(h-1)$$

where h is the number of values left untouched.

References

Dixon and Massey, Sections 16-4, 16-5, pp. 330-332.
Owen, Section 7.2, p. 155.

WILCOXON SIGNED RANK TEST

Data are ordered in absolute value. The signed rank test statistic T is the sum of the ranks of the positive values. Asymptotically, T is normally distributed with

$$\text{mean} = n(n+1)/4, \text{ and}$$
$$\text{variance} = n(n+1)(2n+1)/24$$

References

Dixon and Massey, Section 17-2, pp. 341-342.
Hollander and Wolfe, Chapter 3, pp. 27-32.
Owen, Section 11.1, pp. 325-330.

When there are ties and zero values, the asymptotic distribution has a slightly smaller variance, which can be found in

Cureton, E., The Normal Approximation to the Signed Rank Sampling Distribution when Zero Differences are Present, *J. Am Stat. Assoc.*, *62*, 1068-1104, 1967.

t-TESTS

If you need a reference for the one or two sample t-test, you are not prepared to work with this textbook.

CHI SQUARE ESTIMATION OF CONFIDENCE BOUNDS ON VARIANCE

s^2, the sample variance taken from normally distributed data, is such that, when divided by the underlying variance, it has a chi square distribution with $(n-1)$ degrees of freedom. To get confidence bounds on the underlying variance, divide the sample variance by the appropriate bounds from a chi square table.

POWER TRANSFORMATIONS

The random variable Y is a power transformation of X if

$$Y = (x^t - 1)/t \quad \text{if} \quad t \neq 1, \; \log(X) \text{ otherwise}$$

If there is a known relationship between the variance and the expectation of X, then a power t can be found that stabilizes the variance of Y for many such relationships. Frequently, the variance stabilizing transformation is also a normalizing transformation.

References

Dixon and Massey, Chapter 16, pp. 322-327.
Kendall and Stuart, Section 31.10, p. 469.

CONFIDENCE BOUNDS ON $E(X)-E(Y)$ BASED ON WILCOXON TESTS

If $(x(1), x(2), \ldots, x(n))$ is one sample of data and if $(y(1), y(2), \ldots, y(m))$ is the second sample of data, a fixed value can be added to each of the $y(i)$ that will cause the difference between the two samples to be barely significant with a Wilcoxon test. Similarly a fixed value can be subtracted from each $y(i)$ that will cause significance to be barely reached in the other direction. These two values are bounds on the difference with coverage greater than or equal to the complement of the rejection probabilities used. There is a less tedious way of computing those bounds by ordering all possible pairs of differences $(x(i)-y(j))$ and following

Hollander and Wolfe, Section 4.3, pp. 78-81.

The following are examples of Exercises 2.2-2.6 partially worked out:

Exercise 2.2

TABLE 2.1

Starting at row 4, column 1

| Set | Number of Values | |
	Above 65	Below 51
1	0	5
2	0	6
3	1	6
4	0	8
5	3	6
6	1	5
7	0	6
8	1	4
9	0	4
10	0	7

TABLE 2.3

Starting at row 5, column 6

| Set | Number of Values | |
	Above 65	Below 51
1	4	4
2	5	4
3	3	4
4	3	4
5	0	6
6	1	5
7	1	6
8	3	5
9	6	2
10	1	4

TABLE 2.2

Starting at row 6, column 5

| Set | Number of Values | |
	Above 65	Below 51
1	0	4
2	1	6
3	1	3
4	1	4
5	0	6
6	1	4
7	2	4
8	0	5
9	0	6
10	0	5

TABLE 2.4

Starting at row 3, column 3

| Set | Number of Values | |
	Above 65	Below 51
1	2	7
2	2	6
3	3	6
4	4	4
5	3	6
6	3	6
7	1	6
8	2	5
9	3	4
10	1	8

Insights

1. A test for assymmetry based on the number of values less than
   the mean would be more powerful than a test based on number
   of values more than 2 SD above the mean.
2. Tests based on either set are sufficiently powerful to detect
   the degrees of assymmetry in Tables 2.3 and 2.4.

Exercise 2.3

Random starting points:

Table	Row	Col.
2.1	1	4
2.2	4	8
2.3	3	7
2.4	4	4

Part I

Table	1st #	50th #	Sample Size	Mean	Median	75% Trimmed Mean
2.1	41.4	53.5	10	48.37	42.1	44.45
			30	48.46	48.3	47.17
			50	48.63	47.3	45.28
2.2	60.1	38.1	10	49.94	47.9	50.12
			30	51.48	50.95	51.66
			50	51.07	50.95	51.10
2.3	52.1	62.1	10	55.75	57.65	55.70
			30	54.92	50.5	52.70
			50	55.20	50.25	52.75
2.4	45.1	45.6	10	47.12	44.05	46.82
			30	45.68	47.0	44.79
			50	47.33	46.4	44.59

Note how all three measures of central tendency are not affected
much by the increasing assymmetry when going from Table 2.1 to 2.2
to 2.3 to 2.4.

Exercise 2.3: Part II

Example of calculations using data from Table 2.3

Observed

x	x-50	x-60	x-70
52.1	2.1	- 7.9	-17.9
82.5	32.5	22.5	12.5
34.4	-15.6	-25.6	-35.6
19.1	-30.9	-40.9	-50.9
63.2	13.2	3.2	- 6.8
27.0	-23.0	-33.0	-43.0
70.1	20.1	10.1	0.1
70.5	20.5	10.5	0.5
43.9	- 6.1	-16.1	-26.1
94.7	44.7	34.7	24.7
44.3	- 5.7	-15.7	-25.7
82.1	32.1	22.1	12.1
40.4	- 9.6	-19.6	-29.6
35.6	-14.4	-24.4	-34.4
45.7	- 4.3	-14.3	-24.3
76.5	26.5	16.5	6.5
34.7	-15.3	-25.3	-35.3
33.8	-16.2	-26.2	-36.2
41.1	- 8.9	-18.9	-28.9
58.4	8.4	- 1.6	-11.6
21.1	-28.9	-38.9	-48.9
85.5	35.5	25.5	15.5
48.9	- 1.1	-11.1	-21.1
55.2	5.2	- 4.8	-14.8
22.8	-27.2	-37.2	-47.2
101.6	51.6	41.6	31.6
39.2	-10.8	-20.8	-30.8
66.5	16.5	6.5	- 3.5
68.1	18.1	8.1	- 1.9
88.6	38.6	28.6	18.6

t tests

first 10 values
mean = 55.57
s^2 = 610.112
S.E. of the mean = 7.811

m	(mean-m)/SE
50	0.736
60	-0.544
70	-1.824

first 30 values
mean = 34.92
s^2 = 536.464
S.E. of the mean = 4.229

m	(mean-m)/SE
50	1.163
60	-1.201
70	-3.566

Wilcoxon Signed Rank Tests

	$m=50$	$m=60$	$m=70$
First 10 values:			
Rank Sum of Pos			
residuals = W	21.0	32.0	34.0
$z=(W-E(W))/$SD	- 0.66	0.46	0.66
First 30 values:			
Rank Sum of Pos			
residuals = W	185.0	289.0	353.0
$z=(W-E(W))/$SD	- 0.98	1.16	2.48

3

PAIRED DATA

INTRODUCTION

There are 27 tables (Tables 3.1-3.27) of paired values (labeled X and Y) in this chapter. The relationships between the paired values that can affect the general patterns of data can be thought of in terms of mean differences, relative variances, and correlations. These 27 tables give examples of all combinations of the following conditions:

1. $E(Y)-E(X)$ = Delta = 0, 1, or 4 standard deviations;
2. $Var(Y)/Var(X)$ = Variance Ratio = 1, 2, or 5;
3. Correlation (X,Y) = 0, 0.5, or 0.9.

With the exercises in the chapter, the student should gain a "feel" for how well-behaved paired data look, when there is no underlying differences between X and Y (Delta = 0, Variance Ratio = 1) and also how mean shifts and differences in variance can affect those patterns. Finally, by introducing zero, slight, and extreme correlation, the data display the effects of having additional information about Y contained in the observed values of X.

It is true, of course, that the first two moments do not cover all the ways in which paired random variables can be related. In fact, many of the clever counter-examples that have been concocted to disparage linear regression involve random variables that match

on the first two moments but differ on the higher order moments. No
attempt is made in this book to display such complex relationships.
The purpose of this book is to provide the student with some feel
for the relative differences in random patterns that can be dis-
cerned by examining the fall of data with an experienced eye. The
author does not understand, himself, how such more complex relationships will appear, if they appear at all, in the random patterns of
data.

Since there are 27 tables in this chapter, the exercises are
constructed in a generic form, each exercise describing things to
do with a given table. It would be foolish for the student to
attempt each of these exercises 27 times. The student should treat
this chapter as he or she might treat the exercises in a classical
mathematics textbook such as Hall and Knight's *Higher Algebra*, that
is, like a woodcutter who does not attempt to chop down every tree
in the forest. He examines most of the trees to get some idea of
their size and difficulty, but he uses his axe on only a selected
few.

TABLE 3.1 Delta=0,   Variance Ratio=1,   Correlation=0

X	Y	X	Y	X	Y	X	Y	X	Y
11.7	10.9	7.0	9.3	8.9	8.4	10.5	10.8	10.8	9.6
10.8	9.6	10.2	12.4	9.8	8.4	11.8	11.6	8.1	10.3
10.8	8.8	8.2	11.5	8.5	9.3	11.8	9.8	8.0	10.7
9.7	9.1	9.0	12.1	10.1	11.4	9.2	10.8	10.9	12.8
8.6	10.2	10.2	11.0	10.0	11.2	9.1	10.0	10.4	8.6
11.4	9.9	9.7	9.4	10.4	11.4	10.7	9.7	10.7	11.2
10.7	10.9	8.7	10.3	11.5	9.5	10.4	10.8	8.7	9.6
11.9	9.7	8.6	9.1	9.3	7.8	8.8	12.2	8.3	9.8
10.3	10.9	10.8	8.4	11.1	10.0	11.9	10.2	9.0	9.2
11.9	10.3	12.2	11.2	10.3	7.0	10.1	11.3	10.7	9.8
7.9	9.4	8.2	8.0	10.6	9.9	7.7	9.8	9.2	8.7
11.3	11.3	12.1	6.9	9.6	8.6	8.9	9.3	12.2	9.6
10.6	10.7	9.7	10.8	6.9	9.5	9.9	10.8	8.6	10.3
7.1	11.3	9.3	11.5	10.3	10.6	10.6	11.4	11.3	11.4
10.9	7.7	11.3	7.5	9.5	9.2	10.3	12.0	9.4	11.8
10.6	8.7	7.7	11.2	9.4	8.9	12.0	9.0	10.9	11.4
12.2	8.6	6.6	10.8	10.3	10.0	10.2	11.7	8.1	9.1
11.5	9.6	9.8	11.3	8.0	10.6	9.5	11.0	12.4	7.7
10.3	9.5	9.8	10.0	9.5	8.7	10.8	10.4	10.3	10.0
10.6	10.2	8.5	12.1	10.5	9.6	11.0	10.7	11.0	11.0
9.2	9.9	10.6	7.0	10.8	12.5	12.6	8.3	9.1	9.0
10.3	9.4	8.5	9.5	10.3	11.1	9.1	7.8	11.1	9.7
9.2	10.4	9.2	9.6	10.0	7.8	10.7	9.3	9.5	7.1
9.8	10.4	10.3	8.2	10.8	8.9	8.0	11.1	11.6	10.6
9.0	8.6	9.6	12.4	12.5	10.3	10.4	9.4	10.7	8.7

TABLE 3.2 Delta=0,   Variance Ratio=1,   Correlation=0.5

X	Y	X	Y	X	Y	X	Y	X	Y
8.3	12.1	8.6	8.0	10.2	12.2	9.9	10.4	9.7	9.4
9.7	9.5	10.1	10.4	10.0	12.2	10.0	10.9	11.5	10.9
8.3	9.7	10.4	9.4	9.0	8.8	10.8	10.5	8.3	10.7
8.6	9.8	9.5	7.9	9.6	11.3	11.3	10.7	10.0	10.3
9.8	10.6	9.6	9.7	10.6	12.2	11.2	10.2	10.3	9.3
9.9	10.4	8.2	8.8	9.1	9.1	11.8	10.2	7.5	10.3
9.3	9.5	8.4	8.6	10.5	8.5	9.7	11.6	9.3	10.0
9.5	12.4	10.1	10.9	12.6	12.7	10.8	13.4	10.0	10.6
10.2	9.9	10.0	10.4	11.5	10.1	8.3	8.9	10.2	7.5
9.3	11.1	9.4	7.9	10.5	10.6	9.4	10.9	8.7	9.5
8.2	10.0	8.1	7.7	10.7	10.6	9.3	9.8	10.9	11.0
8.3	8.5	10.3	9.3	10.5	9.4	10.4	10.9	8.8	9.0
9.5	8.9	8.2	10.5	9.5	10.9	9.9	8.8	10.8	9.7
12.1	10.8	9.5	10.4	9.8	10.7	9.7	10.0	10.2	11.3
9.6	10.6	10.6	10.0	9.3	10.0	8.7	9.4	9.3	9.2
12.0	11.7	10.8	12.4	10.4	9.8	11.3	11.0	8.5	10.0
11.6	12.7	10.7	9.7	8.5	9.4	7.2	10.2	11.1	11.0
9.3	11.1	8.3	10.5	11.3	9.7	9.2	9.9	11.9	9.7
8.2	9.5	9.9	9.4	10.5	11.6	9.2	10.3	8.2	8.5
9.1	10.5	12.1	10.8	8.0	8.8	9.5	9.5	8.3	9.0
8.8	9.3	10.4	10.4	8.4	6.9	10.0	10.4	11.3	11.9
11.2	8.2	11.1	8.6	8.8	9.2	12.5	10.1	8.3	10.9
8.4	11.8	10.7	9.1	10.1	8.6	9.5	10.3	10.2	8.3
9.0	8.9	8.0	8.8	9.9	9.1	8.6	8.2	9.2	10.6
9.2	11.1	10.0	10.0	9.5	10.1	10.4	9.2	11.2	9.8

TABLE 3.3 Delta=0,   Variance Ratio=1,   Correlation=0.9

X	Y	X	Y	X	Y	X	Y	X	Y
7.5	7.7	10.0	10.5	10.5	10.3	10.8	10.1	10.2	10.0
10.1	10.3	11.2	11.4	9.9	9.4	8.7	9.3	10.0	9.6
9.8	9.9	11.4	9.8	10.0	10.0	9.5	9.5	10.6	10.2
11.5	11.3	9.2	10.3	9.0	9.1	9.9	9.5	10.0	9.6
9.9	10.5	8.9	8.7	11.6	12.1	10.2	9.8	10.3	10.0
8.4	7.7	8.8	8.1	9.7	9.9	10.8	10.8	9.6	9.5
12.6	12.1	8.3	8.7	10.7	10.2	9.5	9.5	10.6	11.8
10.9	10.9	9.8	10.2	10.2	9.8	7.8	9.0	10.3	10.9
10.4	10.9	10.5	11.1	9.7	10.2	8.3	8.9	11.6	11.5
9.9	8.5	12.0	11.8	10.8	10.6	11.2	9.5	9.6	9.9
10.2	9.5	11.2	10.4	9.2	9.0	9.7	10.1	8.5	8.1
12.2	12.5	9.4	10.2	11.0	10.5	9.5	9.8	9.2	9.2
10.3	10.2	11.0	11.8	10.6	9.9	10.8	10.2	8.8	8.3
10.1	9.9	9.3	9.4	8.9	8.6	9.4	9.4	10.4	10.1
11.4	11.9	9.9	10.3	8.1	8.6	10.5	9.8	8.7	9.1
12.8	12.9	10.9	10.8	11.9	11.1	10.4	10.3	9.1	9.2
8.7	9.6	11.4	9.4	12.1	10.3	10.4	11.2	8.9	9.2
9.4	10.6	11.6	10.6	8.9	7.8	11.0	12.5	10.2	10.1
12.0	11.9	9.3	9.1	9.6	10.6	0.0	0.7	10.4	10.1
9.3	9.2	11.0	10.8	9.0	9.2	9.6	9.7	10.0	10.4
8.6	8.0	12.2	11.2	11.0	10.8	10.7	10.7	10.5	11.2
10.9	10.3	10.0	10.0	8.5	9.4	11.7	12.2	12.0	12.1
10.3	10.7	10.0	10.3	11.7	11.2	10.5	10.1	9.4	9.5
9.7	10.0	9.0	10.0	8.4	9.5	12.2	11.3	8.9	9.6
10.4	9.9	8.9	9.6	10.5	10.4	10.3	11.8	9.1	8.9

TABLE 3.4   Delta=0,   Variance Ratio=2,   Correlation=0

X	Y	X	Y	X	Y	X	Y	X	Y
9.4	13.1	10.1	9.2	9.3	10.1	9.1	9.2	9.4	8.7
10.5	8.1	11.7	12.8	11.0	12.5	9.2	6.0	9.3	9.3
9.6	11.0	9.7	9.2	8.9	12.5	7.6	8.8	8.8	13.7
9.8	7.5	11.5	7.9	10.9	13.1	10.6	10.2	12.3	5.7
8.5	12.6	8.0	7.8	12.2	4.0	14.1	11.8	11.7	10.0
8.4	11.6	11.9	5.4	10.8	5.7	10.5	9.9	11.2	11.4
9.2	9.1	8.5	7.6	8.2	13.9	10.0	7.8	11.0	8.9
9.6	6.1	10.7	7.0	9.6	12.4	7.7	8.7	6.1	8.1
8.5	9.4	11.6	3.7	10.5	11.2	9.6	4.8	11.5	10.1
10.3	9.0	9.2	9.1	6.7	8.4	10.3	8.3	10.1	12.2
11.1	10.4	11.4	12.6	11.0	11.5	9.3	12.9	8.2	12.0
11.7	9.2	10.5	6.7	11.4	11.1	10.2	11.4	9.4	12.9
10.4	11.7	8.0	11.9	8.8	10.4	12.6	10.4	9.7	8.5
11.7	6.4	11.4	8.4	10.1	9.0	11.9	11.5	9.0	11.6
8.9	8.9	8.1	6.1	12.1	9.3	10.5	13.2	10.4	11.1
9.3	11.4	12.1	8.7	8.2	9.4	11.1	11.6	12.8	8.6
7.9	11.2	8.8	9.6	9.5	6.0	8.9	12.4	9.9	10.4
11.4	8.0	8.7	11.8	11.0	11.5	8.0	10.9	11.6	10.0
9.2	10.8	10.3	10.0	10.7	13.6	10.0	9.1	11.5	13.9
9.6	8.3	9.7	7.9	11.5	10.3	8.0	8.7	9.9	11.3
7.9	7.6	11.4	12.9	9.3	11.8	10.0	14.0	8.5	7.3
10.0	6.2	10.2	11.6	10.4	9.4	7.5	7.4	9.2	11.5
10.6	9.0	10.7	11.0	9.5	11.8	10.1	11.9	11.8	12.8
8.9	12.3	10.7	9.0	10.9	8.7	7.1	10.2	10.1	11.6
10.1	8.8	9.0	13.1	10.3	8.7	10.2	11.3	11.8	7.6

TABLE 3.5   Delta=0,   Variance Ratio=2,   Correlation=0.5

X	Y	X	Y	X	Y	X	Y	X	Y
8.8	7.5	11.9	8.1	10.0	11.1	9.9	9.8	11.1	10.4
9.5	9.8	8.0	8.5	9.9	12.9	9.6	9.2	8.7	8.3
10.6	8.3	11.0	10.6	10.1	10.1	9.5	10.3	9.4	8.8
11.3	13.1	10.5	8.6	9.4	7.6	9.4	6.3	9.3	10.3
9.1	10.7	10.0	4.7	9.2	8.1	9.5	9.6	7.4	12.2
10.5	9.6	10.5	8.5	10.8	12.3	11.2	11.2	10.5	11.1
9.3	8.4	10.0	8.3	10.9	9.4	11.4	10.3	11.4	9.8
10.8	11.8	9.0	10.1	10.0	12.4	9.8	9.7	8.2	7.2
9.4	10.8	9.1	6.9	8.1	10.5	9.1	9.7	10.4	9.8
9.6	11.9	10.8	13.0	11.1	11.2	10.6	11.5	8.8	6.5
8.4	6.9	10.3	15.4	10.3	9.0	9.1	9.1	10.0	11.6
10.9	9.0	9.6	10.1	11.1	11.5	11.4	9.5	7.9	6.5
9.9	8.6	10.1	8.8	10.9	10.4	11.5	9.3	10.8	8.6
7.2	10.0	11.3	12.0	8.6	9.8	10.6	11.3	11.5	10.6
8.6	9.2	9.2	7.7	12.7	10.6	11.4	9.9	10.9	7.5
9.0	9.1	7.3	10.9	7.2	5.9	11.9	7.0	8.2	10.9
10.4	10.0	10.1	8.5	9.8	11.5	10.5	9.7	10.4	9.3
8.1	8.7	8.8	9.2	10.2	15.0	10.1	11.0	8.6	10.6
9.4	8.4	10.5	9.4	8.9	8.7	10.0	6.9	8.8	8.2
9.1	9.4	10.6	9.4	10.2	9.5	9.6	9.9	11.0	12.6
10.6	10.1	9.3	8.2	11.4	9.2	11.3	13.0	7.9	8.5
12.0	13.1	10.1	10.6	11.4	8.2	9.7	9.5	7.5	10.0
11.6	9.9	11.4	11.6	11.4	9.4	10.6	11.4	11.3	8.5
9.0	11.5	7.6	8.6	9.3	8.0	11.2	9.7	11.0	10.9
12.4	10.1	11.3	10.0	7.4	8.5	10.5	10.1	10.5	10.9

TABLE 3.6 Delta=0,   Variance Ratio=2,   Correlation=0.9

X	Y	X	Y	X	Y	X	Y	X	Y
11.1	11.5	12.3	12.9	9.4	8.7	7.1	7.0	11.8	11.6
10.5	11.5	9.9	9.7	7.7	5.6	9.0	9.7	10.3	10.5
9.6	9.8	9.8	8.9	12.4	14.0	12.6	13.0	11.3	11.9
9.9	10.2	8.4	7.9	10.1	9.5	11.7	13.2	9.6	7.9
11.2	12.0	9.2	9.5	10.3	10.0	7.3	6.9	8.4	7.4
10.1	10.4	10.7	11.8	11.2	11.9	10.8	11.9	10.1	10.8
8.2	6.8	10.6	8.9	11.2	11.0	8.1	7.6	9.4	7.9
10.5	10.3	9.0	8.7	10.2	10.5	10.2	10.0	10.2	10.4
11.3	10.4	9.8	9.6	10.2	9.8	10.8	10.7	8.6	7.3
9.9	9.9	9.5	9.7	10.6	11.1	9.1	7.8	10.9	11.4
10.2	10.7	9.3	8.8	11.5	12.1	10.7	10.9	10.3	10.1
9.3	9.0	10.7	11.6	10.1	9.1	9.9	9.1	9.9	7.7
8.5	8.0	11.2	11.1	8.5	9.0	9.1	9.9	6.8	7.9
9.8	9.7	11.0	12.4	10.6	11.5	11.4	13.2	10.8	10.9
9.4	8.0	12.2	11.5	11.1	11.6	9.9	11.3	10.5	9.8
10.5	10.2	10.4	12.1	11.1	10.5	10.6	11.8	9.7	10.9
9.9	11.5	9.5	9.5	9.3	9.6	9.3	8.1	9.0	9.2
9.4	9.9	11.2	11.6	9.8	9.9	10.2	11.1	10.5	9.9
9.7	9.0	9.8	8.2	9.9	11.1	9.2	9.1	9.3	8.5
10.0	9.4	10.6	11.0	11.0	11.2	9.1	10.2	8.0	9.5
10.2	10.5	10.6	10.7	12.7	13.7	10.0	9.3	9.4	9.8
11.0	11.5	9.9	10.1	8.8	8.9	11.1	11.4	9.6	8.4
10.0	11.2	9.9	11.7	8.7	8.0	11.8	12.9	11.3	10.5
11.2	11.5	7.9	9.0	11.0	12.5	9.7	8.7	11.4	12.3
9.6	10.2	8.5	8.7	9.9	11.5	10.1	9.9	9.9	9.6

TABLE 3.7 Delta=0,   Variance Ratio=5,   Correlation=0

X	Y	X	Y	X	Y	X	Y	X	Y
6.9	8.8	8.4	9.1	6.9	11.4	10.7	11.3	9.3	4.0
11.1	13.5	12.1	14.3	9.8	2.4	8.2	13.8	10.9	10.4
10.2	8.2	9.9	10.0	9.5	10.8	9.1	5.7	9.8	14.1
12.1	17.6	9.9	14.7	10.5	7.1	10.2	13.5	10.4	16.8
7.3	8.2	10.7	12.7	9.3	13.0	8.8	13.0	12.0	11.3
9.1	9.6	11.1	8.5	9.7	12.1	9.1	12.3	11.6	10.7
9.3	9.0	8.7	7.1	10.4	11.3	8.5	9.7	9.8	7.9
9.9	13.1	9.5	17.8	9.5	12.6	12.6	8.6	8.1	8.1
10.9	12.0	9.3	8.9	10.6	9.9	11.2	6.6	10.3	15.1
5.9	11.5	11.3	10.4	10.8	10.8	10.2	9.8	9.4	12.8
10.7	13.0	11.8	11.7	8.1	15.7	8.4	12.9	7.7	10.3
10.2	10.9	8.0	13.2	8.4	10.0	10.0	12.8	10.1	10.7
10.0	9.0	9.0	13.8	11.5	12.5	11.7	6.6	11.5	10.3
10.9	6.6	10.9	5.7	11.1	7.6	10.0	12.9	9.2	10.6
10.3	9.6	10.4	8.9	12.1	10.4	13.5	8.1	10.4	14.3
10.2	11.8	8.4	10.0	8.9	11.8	11.3	9.4	8.8	8.7
9.2	9.3	11.9	4.2	9.7	13.2	8.5	15.0	11.1	11.3
10.9	11.7	12.9	8.8	9.6	3.5	10.4	13.1	11.2	10.9
8.7	5.7	10.1	7.5	9.1	8.7	10.9	10.5	10.4	15.3
11.0	4.5	11.8	8.1	8.0	8.6	10.7	10.1	9.1	13.6
9.1	11.8	9.7	5.9	10.4	5.9	11.7	5.1	10.0	15.6
10.6	11.0	11.3	15.5	9.8	12.2	10.2	6.6	12.3	10.3
9.1	6.4	12.4	15.2	11.4	10.8	10.0	10.1	10.7	12.5
8.7	13.8	12.5	10.5	11.1	14.1	9.2	12.3	10.4	11.1
7.2	10.7	8.2	8.7	11.6	10.3	12.0	6.3	12.0	8.8

TABLE 3.8 Delta=0,   Variance Ratio=5,   Correlation=0.5

X	Y	X	Y	X	Y	X	Y	X	Y
9.2	10.6	10.3	14.0	10.6	12.8	11.4	9.5	8.5	8.2
10.3	6.9	11.9	8.4	7.5	12.4	7.9	8.9	9.9	8.4
9.8	6.4	7.8	8.7	8.8	7.1	12.0	8.4	9.6	7.4
9.5	12.7	10.9	7.0	12.1	8.6	11.0	14.9	7.7	10.5
10.1	5.9	11.4	8.0	10.4	11.0	10.4	14.4	8.1	7.9
7.8	8.2	11.5	7.9	10.2	10.2	8.8	7.9	11.4	7.6
10.2	12.6	10.6	9.3	10.6	10.0	9.8	13.3	8.6	8.3
10.5	9.9	9.9	9.9	8.9	7.3	10.7	12.7	9.8	11.8
9.0	7.8	8.3	8.0	11.2	12.5	9.2	8.9	9.8	9.7
7.5	8.7	9.3	16.1	9.4	9.5	10.4	4.8	9.1	10.6
9.7	8.1	12.4	13.6	12.2	18.5	9.5	9.5	12.8	11.8
10.1	6.8	10.4	6.3	11.1	7.3	8.9	11.4	9.1	9.7
9.9	7.2	10.2	6.3	9.1	10.4	10.6	10.0	10.4	9.1
11.3	12.4	11.5	4.3	11.0	8.0	11.4	11.8	9.4	8.7
10.4	8.9	10.7	6.4	9.2	12.4	10.3	9.8	9.6	8.9
12.5	15.3	11.0	8.2	8.4	10.2	9.9	10.9	11.7	12.0
12.2	9.1	11.0	9.9	12.0	13.0	10.5	9.9	9.2	6.7
11.3	15.3	9.4	4.8	11.8	13.6	9.4	8.3	7.2	6.6
9.8	10.3	10.0	11.3	9.3	8.4	8.6	8.3	11.1	7.9
9.5	10.0	10.8	16.3	9.2	9.4	9.5	12.0	9.1	10.4
11.7	9.9	10.6	8.2	9.6	12.4	10.9	13.5	10.3	12.9
9.8	6.4	10.3	9.5	8.8	6.6	11.4	11.1	9.9	8.6
10.3	10.5	10.6	14.9	11.0	10.3	11.9	9.9	10.8	6.3
12.6	11.3	8.4	10.7	9.5	8.7	9.5	11.5	12.7	8.2
12.2	10.0	11.2	13.8	10.2	7.6	7.7	13.1	9.2	9.7

TABLE 3.9 Delta=0,   Variance Ratio=5,   Correlation=0.9

X	Y	X	Y	X	Y	X	Y	X	Y
7.5	5.2	9.7	7.7	10.8	12.3	9.2	8.1	11.2	12.7
7.9	5.1	11.8	12.2	10.6	12.6	8.5	7.8	9.0	8.9
11.4	13.3	9.2	6.3	10.0	8.8	9.8	9.7	8.4	6.8
10.2	10.9	8.9	8.4	10.9	13.6	12.1	12.7	9.5	8.8
10.4	11.5	9.5	8.9	12.5	16.6	11.5	12.9	8.3	8.7
9.5	7.8	10.8	10.8	10.2	13.5	9.4	9.7	11.7	11.4
10.6	10.0	8.6	4.2	10.5	12.6	10.9	12.0	10.2	9.4
9.7	9.1	9.7	9.7	11.4	13.8	9.3	8.8	8.3	5.9
9.0	8.9	10.3	9.9	10.3	9.7	11.1	13.7	11.6	10.9
8.7	5.4	11.0	13.7	9.2	8.3	9.4	10.5	10.2	11.1
9.4	10.0	10.0	7.8	10.6	11.5	9.4	8.2	9.6	7.9
10.5	9.6	11.2	13.1	11.2	11.8	10.7	12.4	9.9	9.6
12.1	15.7	10.1	9.6	9.3	8.3	8.0	5.3	12.2	12.0
11.0	10.6	10.2	8.9	10.9	13.2	9.9	7.8	9.5	9.8
10.2	12.2	9.6	7.6	9.7	8.6	10.0	8.1	9.1	11.1
10.7	12.9	11.6	13.2	10.9	13.7	11.1	13.3	11.0	12.6
9.0	11.5	10.7	8.8	10.6	8.1	9.5	9.3	10.5	10.8
8.9	7.9	9.2	5.8	9.4	7.5	10.5	11.9	9.4	5.6
9.5	10.7	10.3	8.6	9.2	7.5	9.9	11.6	10.2	9.5
9.3	7.0	10.8	12.0	11.0	11.5	9.5	8.0	10.7	12.7
10.9	9.4	10.1	10.0	9.7	9.8	11.0	14.9	8.8	10.4
9.4	10.2	8.5	4.7	9.7	8.7	7.7	5.0	8.8	6.8
9.9	7.5	11.1	10.7	11.1	12.0	10.2	10.9	10.6	10.1
10.4	9.9	9.8	9.7	9.7	10.3	9.3	6.3	11.3	11.8
9.3	10.7	8.9	8.4	9.5	8.7	12.8	15.9	9.2	7.9

TABLE 3.10 Delta=1 S.D., Variance Ratio=1, Correlation=0

X	Y	X	Y	X	Y	X	Y	X	Y
9.5	11.3	9.7	8.9	12.2	10.6	7.5	8.2	9.8	14.9
10.8	12.2	9.1	8.9	9.5	11.7	11.3	12.8	6.8	13.2
9.0	10.3	11.7	10.8	12.4	12.2	9.3	10.0	8.8	10.3
10.1	10.9	9.6	9.3	11.1	10.5	10.3	12.7	11.6	11.3
11.4	9.7	10.4	12.6	10.2	10.6	10.0	10.0	7.7	9.0
9.8	8.9	11.2	9.9	9.9	9.7	11.1	7.9	10.5	11.1
8.9	9.5	12.2	8.4	11.4	13.1	9.4	9.4	11.5	11.2
9.5	10.7	8.7	11.2	10.8	10.5	10.2	10.2	9.3	10.8
10.2	11.8	10.2	12.9	12.7	12.6	6.9	11.7	8.8	10.9
10.1	9.0	12.0	10.8	8.2	13.5	6.6	13.7	11.0	13.1
12.2	9.3	12.3	11.0	10.4	12.3	11.3	11.1	7.6	11.1
11.7	9.3	11.0	11.1	11.1	10.7	10.6	9.4	10.4	9.7
9.1	11.3	10.0	8.6	9.9	13.7	11.2	10.0	13.0	10.1
10.5	10.8	9.4	10.3	9.0	11.9	10.3	12.2	10.4	8.6
8.8	9.4	8.4	8.7	8.4	7.7	8.7	11.3	7.9	11.5
13.0	12.0	10.3	9.3	8.7	11.7	8.8	11.7	9.8	12.6
9.8	10.6	10.4	13.5	8.5	8.9	9.5	9.0	9.4	14.2
10.1	8.9	11.8	11.3	9.6	10.8	12.2	11.6	11.4	9.7
9.5	13.5	11.8	10.5	11.2	9.3	9.8	9.8	10.8	11.0
9.1	11.3	11.5	10.4	8.6	13.0	10.2	11.1	9.1	11.0
7.6	13.7	10.6	11.5	10.3	11.9	12.5	11.9	10.3	11.9
6.5	8.3	8.5	12.5	8.0	9.7	7.1	13.7	10.6	10.0
10.4	10.0	10.7	11.8	6.8	10.9	7.9	11.5	11.7	8.4
10.1	10.5	9.8	10.1	9.7	11.7	8.7	10.9	9.4	11.0
9.1	11.8	11.6	11.6	9.4	12.8	11.0	9.3	10.9	9.9

TABLE 3.11 Delta=1 S.D., Variance Ratio=1, Correlation=0.5

X	Y	X	Y	X	Y	X	Y	X	Y
10.5	12.1	10.3	13.1	9.5	10.0	9.7	10.7	12.8	11.9
9.2	12.2	9.5	10.2	11.5	13.9	10.0	9.9	11.5	10.4
9.1	9.3	11.2	13.4	10.2	10.6	10.8	9.7	9.3	9.0
10.6	9.4	10.0	10.4	11.6	9.1	11.1	11.6	9.6	9.6
9.0	11.4	8.4	9.9	10.4	8.9	9.1	11.2	7.3	10.6
8.1	11.4	10.6	12.2	9.7	9.2	7.9	10.6	11.2	11.1
9.6	10.0	10.5	11.3	6.2	9.1	8.9	12.1	9.2	12.0
10.4	10.9	10.8	8.6	11.3	9.5	9.9	11.0	10.1	12.7
10.1	11.5	10.3	11.0	9.6	9.0	9.4	10.9	9.3	7.8
9.9	12.7	10.2	10.3	13.1	9.8	13.0	13.3	12.0	10.9
11.9	13.9	8.2	10.4	10.1	12.3	10.6	12.2	10.7	13.1
8.6	11.2	10.2	12.3	10.1	11.3	11.0	12.0	10.1	10.0
10.3	10.3	7.6	10.3	9.9	9.4	9.8	8.4	9.0	9.9
10.2	10.0	12.3	9.4	8.8	12.5	11.0	10.9	7.5	9.3
10.3	10.5	9.2	12.8	10.0	9.9	8.6	10.3	9.0	11.2
9.4	7.6	10.5	11.9	7.2	9.7	7.0	9.9	10.6	13.1
11.8	12.7	9.9	10.6	11.6	14.2	9.9	10.2	10.3	11.9
9.2	10.1	10.1	12.8	8.8	13.7	9.6	10.0	0.8	10.3
10.2	11.9	8.5	11.2	9.0	10.0	10.8	12.4	9.9	12.1
7.4	10.1	10.6	9.9	10.7	12.2	10.9	10.1	11.2	12.7
8.3	9.5	10.8	14.4	10.2	9.9	9.7	10.4	8.5	10.6
9.3	8.7	7.6	10.3	7.7	10.9	11.2	11.6	10.4	12.8
10.8	13.4	12.7	10.3	9.4	9.2	9.3	10.9	9.6	10.7
10.7	11.6	9.2	9.5	9.5	12.2	10.3	12.7	9.5	11.0
12.5	12.7	10.1	11.0	11.9	10.8	9.4	11.3	8.5	12.3

TABLE 3.12 Delta=1 S.D., Variance Ratio=1, Correlation=0.9

X	Y	X	Y	X	Y	X	Y	X	Y
11.2	11.2	10.5	11.8	9.7	10.0	10.5	12.0	12.8	13.2
9.3	11.3	10.1	11.1	8.7	10.2	10.5	12.0	12.0	13.0
7.7	9.1	8.9	10.0	10.1	10.2	10.6	11.8	10.6	11.5
10.5	11.7	8.2	10.2	9.4	11.0	9.8	10.4	11.3	12.6
9.5	11.0	8.8	10.1	11.4	11.5	11.2	12.7	9.4	9.8
8.4	9.7	9.2	9.7	8.8	10.4	10.6	11.8	8.8	10.2
10.8	10.6	11.0	11.7	10.7	11.5	8.2	9.7	11.2	12.3
9.5	11.4	10.7	11.8	10.6	11.1	8.2	9.9	11.3	12.6
10.0	11.8	10.9	12.4	8.6	9.8	10.0	11.0	10.1	12.3
9.3	10.2	12.6	14.2	7.4	8.6	10.3	11.1	10.4	10.8
11.4	11.9	9.6	11.7	11.3	11.8	12.4	11.9	12.1	11.8
9.8	10.1	10.9	11.8	9.1	10.8	11.8	11.6	10.8	11.9
8.2	10.5	8.8	10.0	9.6	10.9	10.4	10.4	10.8	11.6
10.3	12.1	11.4	12.6	11.7	11.7	10.6	11.7	10.5	10.5
10.3	11.4	11.6	14.0	11.2	11.5	12.0	12.4	9.9	11.8
8.9	9.6	9.8	11.0	8.6	10.3	11.9	12.7	8.0	9.3
12.0	12.2	9.7	10.4	10.2	10.6	9.9	11.4	10.3	11.9
10.5	10.5	9.4	10.2	10.8	11.1	8.9	9.6	10.1	11.5
9.7	10.2	10.7	10.9	8.8	8.7	7.6	10.0	9.4	9.8
9.7	11.2	8.0	8.9	10.9	11.9	10.2	10.9	10.5	10.8
7.2	8.6	11.1	13.6	9.3	10.6	10.5	11.7	9.6	9.8
9.7	11.6	10.5	11.7	10.5	11.3	10.3	11.3	11.0	12.7
9.5	10.5	9.8	10.7	9.1	10.0	10.2	11.6	12.3	13.3
11.5	12.2	9.6	10.8	11.0	12.1	8.4	9.1	8.2	8.7
10.4	10.4	8.5	9.3	9.8	11.6	8.8	10.5	8.6	9.6

TABLE 3.13 Delta=1 S.D., Variance Ratio=2, Correlation=0

X	Y	X	Y	X	Y	X	Y	X	Y
10.6	11.8	8.1	11.2	12.4	12.4	10.9	11.8	12.1	12.1
11.1	9.8	9.1	10.1	10.6	11.7	10.7	6.8	7.9	11.6
10.9	13.0	9.0	15.1	9.6	11.0	9.9	6.9	7.6	14.7
9.9	10.7	8.8	9.5	9.8	13.1	9.9	9.9	10.6	13.2
8.3	10.4	11.9	9.4	10.6	9.5	10.2	10.9	12.0	10.1
11.6	9.1	10.6	11.1	9.6	8.9	11.7	14.5	11.3	12.2
10.3	10.6	9.6	9.8	7.9	14.3	9.3	9.3	9.1	11.3
8.7	11.9	9.0	10.8	9.6	9.4	12.3	12.2	9.7	11.0
9.8	10.7	11.0	12.4	8.7	15.1	9.7	10.6	11.3	8.4
10.9	10.6	12.8	9.4	10.8	12.5	8.9	9.0	9.8	14.2
8.5	8.3	12.6	14.1	11.3	9.5	11.4	12.4	10.8	9.6
13.0	14.1	11.3	10.7	9.2	9.7	10.4	9.4	12.9	9.7
9.3	9.9	10.8	7.2	11.0	10.0	10.4	13.2	9.5	9.0
10.3	11.8	9.2	9.8	9.6	10.8	10.3	10.1	11.2	9.5
9.4	12.7	12.2	12.0	9.3	9.7	10.9	9.1	10.8	9.8
10.4	9.1	9.3	11.4	11.9	10.5	9.1	10.5	9.9	11.8
8.8	9.6	10.5	9.8	9.6	11.0	12.9	10.3	5.5	14.3
11.5	12.3	8.5	8.0	10.4	12.1	10.0	11.1	7.8	10.5
11.2	8.5	12.7	12.4	7.9	10.3	11.3	10.4	9.7	13.3
12.4	8.3	9.6	12.0	8.3	10.8	10.0	11.7	7.6	9.1
8.5	14.3	11.2	11.8	12.3	11.3	10.1	10.0	8.2	11.6
7.2	8.4	10.1	10.8	11.7	7.9	10.2	7.3	9.1	8.7
8.8	10.2	11.8	13.3	9.7	11.8	9.8	10.1	11.4	10.4
10.7	9.5	9.2	9.0	11.8	11.6	9.8	11.1	11.8	10.0
10.2	9.1	9.2	8.5	9.9	12.1	12.9	10.0	11.0	9.7

TABLE 3.14 Delta=1 S.D., Variance Ratio=2, Correlation=0.5

X	Y	X	Y	X	Y	X	Y	X	Y
9.6	10.9	8.7	8.6	9.1	13.0	9.4	12.3	8.6	10.4
11.0	12.0	10.3	11.6	10.4	9.5	9.6	11.5	7.8	8.3
12.0	12.0	11.6	12.8	9.0	9.7	8.8	12.3	10.3	13.3
6.6	10.4	9.6	9.9	9.3	10.5	10.9	11.6	8.7	10.1
11.2	14.8	12.0	12.9	11.8	11.1	9.8	12.7	8.8	10.0
9.8	8.5	9.3	10.0	9.0	11.5	10.2	11.2	10.4	13.2
10.0	9.4	10.3	9.5	11.7	11.2	10.0	9.1	9.1	10.9
10.8	11.4	8.2	7.3	8.4	9.4	10.9	10.6	8.6	12.7
9.6	9.0	9.1	14.7	9.9	9.1	9.6	12.6	12.6	15.0
11.7	10.1	10.3	10.7	10.6	10.2	9.8	12.2	11.8	9.8
11.6	10.6	11.0	12.4	10.2	10.8	8.7	8.4	10.2	12.1
9.2	9.1	11.6	15.4	9.1	10.2	10.0	11.2	11.8	10.4
9.0	8.2	10.3	10.4	9.4	9.3	8.6	9.3	9.4	6.3
10.0	10.2	11.0	11.5	10.3	9.5	10.0	11.0	8.8	10.2
8.1	11.2	10.0	9.0	10.4	10.9	11.1	10.0	10.8	9.6
12.1	10.5	8.9	11.5	10.8	11.8	9.1	10.2	11.8	10.3
9.0	13.3	9.2	12.2	10.5	11.7	10.5	10.9	10.6	10.5
11.1	11.3	11.1	11.8	10.6	13.7	10.7	14.4	9.3	13.1
11.1	10.0	10.2	11.0	10.4	11.4	9.9	11.6	10.6	10.7
9.6	11.9	9.6	10.6	10.0	10.8	9.0	11.9	11.9	9.5
9.4	8.1	10.4	9.0	10.2	8.3	11.2	14.3	9.9	9.8
10.7	15.4	9.9	13.2	10.1	12.2	10.3	8.2	9.9	13.4
10.6	13.1	10.9	12.1	9.7	9.9	11.7	14.1	10.6	11.4
10.5	12.0	10.3	10.4	9.0	9.7	9.9	9.8	8.8	7.8
8.2	12.6	9.3	8.4	11.5	13.1	10.9	13.2	11.6	10.8

TABLE 3.15 Delta=1 S.D., Variance Ratio=2, Correlation=0.9

X	Y	X	Y	X	Y	X	Y	X	Y
9.6	10.7	8.5	9.1	9.7	10.7	10.5	11.5	9.2	9.6
9.9	11.5	10.5	9.9	11.0	12.8	9.0	9.4	10.1	11.3
10.5	10.3	10.9	11.7	8.5	10.0	11.9	13.2	9.5	10.3
9.1	9.1	9.1	9.9	8.9	10.3	7.6	9.6	9.7	9.1
9.6	10.5	9.8	10.8	11.2	12.0	10.1	11.8	8.7	9.9
9.1	10.4	9.3	10.8	8.8	9.6	9.6	10.0	11.3	11.7
10.6	11.4	11.3	11.4	8.8	10.3	8.1	8.2	12.3	13.8
10.7	11.5	10.3	10.9	7.9	9.2	9.8	9.9	9.9	10.8
9.1	10.2	9.5	9.3	10.1	12.4	10.6	12.0	9.2	11.3
8.9	9.6	11.3	12.1	10.5	12.4	9.6	11.1	8.4	9.9
10.3	11.9	8.0	8.6	9.7	9.7	10.5	11.7	8.6	9.7
11.4	13.7	9.6	11.1	8.3	9.2	10.6	12.1	11.7	13.8
8.6	9.1	10.1	10.4	10.2	12.4	8.4	8.4	9.0	9.5
9.3	11.9	10.5	12.2	9.1	10.0	7.7	7.7	11.0	10.3
9.6	8.6	10.9	12.1	9.1	9.9	10.5	11.4	9.3	11.1
10.3	11.9	8.6	9.3	11.2	12.1	10.4	10.7	10.4	13.2
9.8	10.4	10.5	11.2	10.5	12.0	10.7	11.3	8.4	8.6
11.0	11.0	12.5	14.9	10.4	11.5	10.4	12.3	8.3	9.3
10.7	11.0	9.5	9.5	9.9	11.1	8.8	9.4	9.7	10.4
9.5	11.3	11.9	12.4	10.8	12.0	10.6	11.4	10.9	12.9
12.3	14.1	11.1	12.6	12.6	14.1	11.1	13.0	11.5	12.0
9.7	10.3	9.5	10.4	10.7	11.9	10.6	13.2	7.8	9.9
8.2	8.6	9.9	11.1	7.6	6.9	9.1	10.4	9.1	11.6
9.1	10.3	11.5	12.8	8.8	10.2	10.1	12.2	9.3	10.2
8.1	8.9	9.1	9.9	10.9	11.5	11.3	12.5	12.6	13.4

TABLE 3.16 Delta=1 S.D., Variance Ratio=5, Correlation=0

X	Y	X	Y	X	Y	X	Y	X	Y
13.3	12.3	9.3	13.1	10.0	5.8	11.8	14.6	9.4	13.5
9.5	9.7	12.8	15.0	10.3	11.1	7.8	13.3	9.9	11.0
12.5	8.1	6.6	11.3	11.5	10.2	9.6	12.2	9.8	11.6
9.8	10.8	7.5	6.2	10.6	14.6	8.3	13.5	8.5	6.9
10.7	13.5	9.7	11.1	10.8	10.6	10.7	10.2	10.7	9.5
10.0	8.1	9.6	15.1	9.2	10.2	8.5	11.6	11.1	10.4
10.1	15.9	10.4	8.0	9.6	13.2	9.2	10.4	9.3	10.7
11.6	9.8	7.6	10.5	9.3	10.2	11.7	11.1	9.7	9.3
12.8	4.4	9.4	9.4	7.2	15.5	9.9	11.2	12.3	22.0
9.2	12.7	8.5	8.0	10.8	8.4	9.3	7.1	9.4	12.3
11.3	4.7	10.8	13.1	12.5	6.8	11.4	9.9	8.6	15.8
9.6	8.4	8.2	6.9	9.8	7.0	11.4	11.6	9.8	13.8
7.6	10.0	8.5	7.5	11.4	7.3	9.4	13.4	10.8	15.2
9.2	12.1	9.4	7.6	8.9	11.7	11.2	10.1	9.5	12.4
10.7	16.0	8.9	9.4	11.0	7.6	11.6	13.4	11.8	13.0
10.5	13.7	11.8	12.2	10.1	13.7	9.9	14.5	13.3	10.6
9.3	15.0	12.1	10.3	8.6	4.7	9.7	16.5	9.1	8.7
9.0	13.7	11.1	11.6	12.5	13.5	11.9	14.9	12.5	9.7
9.0	2.3	12.3	18.3	11.3	14.5	10.9	15.1	7.5	13.7
8.4	11.6	10.1	6.4	14.1	12.1	9.7	11.0	9.2	10.0
9.9	16.4	10.9	11.4	12.1	10.2	10.7	16.9	7.2	9.7
10.1	12.6	13.0	9.7	10.2	8.7	9.3	8.3	11.3	5.9
11.4	12.9	11.6	9.5	9.8	11.0	8.5	13.5	9.2	14.6
11.5	15.4	8.7	9.8	9.1	12.9	9.0	7.6	10.2	10.7
8.9	11.0	9.7	14.4	8.3	18.4	10.7	11.1	9.1	7.8

TABLE 3.17 Delta=1 S.D., Variance Ratio=5, Correlation=0.5

X	Y	X	Y	X	Y	X	Y	X	Y
9.0	9.9	10.1	10.2	8.5	7.9	10.5	16.0	10.3	12.8
9.3	6.9	10.3	6.3	14.1	10.7	10.8	9.5	9.5	4.3
10.6	8.5	9.2	13.9	9.3	11.2	11.2	12.3	10.5	8.5
9.7	8.7	8.1	4.4	11.4	15.3	10.0	13.8	12.8	14.3
10.6	11.3	8.1	10.0	10.1	8.7	10.0	13.2	10.9	9.5
6.6	2.0	11.1	11.7	10.1	10.5	11.5	12.5	12.1	14.8
8.8	13.7	10.7	14.2	9.6	7.5	11.7	7.5	9.5	5.5
11.0	7.9	9.7	8.5	9.7	8.1	11.4	11.6	10.5	10.6
6.9	7.1	10.6	14.2	11.9	13.5	9.9	9.2	12.5	11.3
7.4	9.5	9.9	12.8	7.3	5.4	9.7	10.8	12.1	9.3
8.3	11.0	9.6	6.3	8.8	5.3	11.6	13.5	10.1	10.5
7.4	7.1	10.8	11.1	9.9	13.2	12.7	12.0	10.2	12.1
11.4	13.8	7.9	13.4	10.7	11.5	9.6	6.1	11.8	9.6
9.5	11.3	10.6	16.4	9.0	13.5	7.6	11.3	10.8	14.5
11.3	10.2	8.6	6.4	8.0	10.1	9.2	12.3	9.2	8.6
11.6	9.7	9.4	8.2	7.5	11.4	11.0	11.5	10.2	10.4
10.6	13.6	8.6	8.3	10.7	12.6	11.0	11.4	7.9	12.6
10.0	9.2	10.5	10.6	9.3	11.6	9.0	6.6	10.4	8.0
11.1	13.5	9.7	12.6	9.8	11.5	9.9	9.0	7.3	8.3
9.0	6.5	8.2	8.2	9.4	9.2	9.7	15.1	10.4	10.7
10.6	11.9	8.8	13.3	9.4	10.5	6.8	10.7	9.3	9.5
11.3	7.5	9.9	12.6	10.9	7.4	11.0	7.1	10.3	14.4
8.5	11.5	11.0	9.1	10.8	14.7	10.2	8.6	10.0	15.2
12.0	10.6	9.2	8.7	9.6	14.0	7.9	12.5	9.8	13.3
11.6	11.0	10.8	10.5	9.8	9.3	10.2	11.2	11.5	15.7

TABLE 3.18 Delta=1 S.D., Variance Ratio=5, Correlation=0.9

X	Y	X	Y	X	Y	X	Y	X	Y
10.3	10.9	9.2	13.4	9.5	10.5	10.2	11.1	9.9	11.4
9.3	9.3	9.2	9.5	10.1	11.3	10.2	10.6	9.4	10.6
9.7	9.3	8.7	7.0	10.2	12.2	9.0	8.2	9.9	10.6
10.5	11.8	10.9	10.9	9.8	12.7	10.4	9.6	10.8	14.5
10.5	11.6	10.7	10.0	10.0	11.2	9.9	7.9	10.0	11.7
9.9	11.7	9.3	8.9	9.3	10.5	11.7	14.4	8.7	8.8
10.6	10.6	9.8	11.0	10.2	9.8	7.7	5.8	8.9	11.4
11.2	15.6	9.2	7.7	10.1	9.7	9.1	8.9	9.2	10.5
8.9	9.9	10.5	13.6	11.6	12.9	8.0	8.0	9.7	11.1
10.4	11.4	9.3	9.8	9.4	10.8	10.3	12.7	10.5	12.9
11.3	12.8	10.8	12.7	9.1	11.6	10.3	12.7	10.0	9.9
10.8	11.5	10.8	14.3	9.8	12.6	9.4	10.7	13.9	19.0
9.6	11.2	10.3	10.5	11.5	15.5	9.9	9.1	11.6	13.9
8.4	8.0	9.5	7.4	11.3	12.9	9.8	10.5	10.2	9.4
9.6	9.5	11.1	14.3	9.9	14.0	10.5	11.3	9.8	13.1
9.3	7.3	12.1	12.6	10.4	13.2	10.0	13.5	10.1	12.4
10.8	12.4	10.6	9.2	8.7	11.4	8.6	9.5	8.7	10.2
9.0	8.8	9.5	8.1	9.6	9.1	10.0	11.6	10.0	11.9
9.5	11.2	9.7	12.7	11.8	13.6	9.5	11.0	9.7	13.3
11.9	13.7	10.8	14.2	9.9	8.4	10.7	13.2	11.0	14.0
12.1	16.5	10.0	9.5	8.4	8.4	9.5	11.2	8.8	8.1
10.1	9.9	11.0	13.5	12.1	14.1	12.5	15.1	8.3	7.0
10.6	10.9	9.4	9.6	9.6	10.6	7.5	7.4	8.4	6.6
9.5	9.3	9.5	11.2	8.0	8.3	7.6	4.9	10.4	10.6
8.5	6.8	9.8	12.0	10.8	13.0	10.7	12.4	11.3	14.4

TABLE 3.19 Delta=5 S.D., Variance Ratio=1, Correlation=0

X	Y	X	Y	X	Y	X	Y	X	Y
7.6	13.1	7.6	14.2	8.3	13.2	10.7	13.7	12.0	15.1
9.0	16.1	10.1	12.9	12.7	13.2	9.7	15.6	10.9	13.0
9.1	11.3	9.0	15.5	9.9	14.0	8.6	12.6	6.8	16.3
11.2	13.9	10.3	17.3	8.9	12.1	9.3	15.5	9.6	13.4
8.4	12.8	10.2	13.4	11.5	15.4	10.1	13.9	12.7	15.7
9.7	14.9	10.5	15.0	8.3	15.7	10.0	12.3	8.4	14.3
9.8	11.1	9.8	15.8	12.6	13.8	9.3	14.0	9.9	15.3
8.7	15.9	9.2	13.5	8.1	15.5	10.8	12.1	10.8	12.8
9.4	11.8	10.0	14.8	8.8	16.3	9.3	13.1	8.7	13.8
11.3	16.3	10.7	16.5	10.0	15.1	8.1	11.7	8.5	14.8
9.4	14.0	14.2	12.8	10.5	12.6	8.6	13.6	11.9	13.1
11.8	14.9	7.9	13.7	10.9	14.6	11.2	13.2	8.7	12.9
10.3	13.7	9.8	14.4	11.1	14.5	9.1	15.1	7.6	12.8
9.0	15.0	7.9	14.9	10.2	13.0	9.1	14.5	9.2	17.5
10.2	14.1	9.7	12.6	10.0	14.7	7.6	14.2	9.2	15.8
9.7	10.1	9.2	16.5	9.1	13.9	12.2	11.8	9.4	12.8
7.4	14.7	11.1	9.4	10.9	14.2	10.5	12.2	11.4	15.0
10.7	13.4	8.2	13.0	10.9	14.7	10.6	14.0	7.4	14.4
9.1	13.1	11.6	11.9	8.9	13.7	9.6	13.2	12.0	15.4
9.3	14.7	9.1	13.6	10.6	15.2	11.5	13.2	9.8	10.7
10.2	13.2	9.5	15.5	7.8	13.4	8.0	15.0	13.2	13.5
12.1	13.7	10.4	15.1	10.1	13.7	9.1	14.3	10.1	16.1
7.4	13.9	10.2	13.1	11.6	13.6	10.0	14.8	12.0	14.1
10.5	15.4	12.2	13.3	11.2	12.8	7.8	15.4	9.4	15.7
9.5	13.7	10.2	12.5	8.2	11.7	9.2	12.9	7.6	11.5

TABLE 3.20   Delta=5 S.D., Variance Ratio=1, Correlation=0.5

X	Y	X	Y	X	Y	X	Y	X	Y
10.8	14.1	9.9	14.2	9.5	13.7	13.7	14.8	6.0	12.7
8.0	14.6	8.9	12.8	10.3	13.0	9.3	14.2	6.8	12.2
8.9	12.7	10.2	15.8	11.5	14.5	9.3	13.8	9.9	12.5
8.4	13.3	11.4	12.9	11.7	16.9	10.8	15.0	10.9	15.9
8.6	15.9	11.1	14.6	9.5	15.6	10.8	13.6	11.4	14.0
12.1	14.2	6.4	12.0	9.8	15.5	10.7	15.1	8.4	13.5
10.2	12.5	9.2	14.6	9.9	12.5	11.3	15.4	10.5	14.1
10.1	14.4	11.1	16.5	9.6	14.3	9.0	13.7	10.2	14.0
12.3	13.5	10.1	12.9	12.5	15.8	10.3	14.6	7.5	12.9
9.0	13.0	9.9	11.9	8.8	12.6	10.8	12.4	10.5	16.3
10.6	15.2	9.5	13.4	7.7	13.1	8.4	13.0	10.7	14.6
9.0	13.8	8.7	15.4	10.7	12.7	10.0	12.8	9.5	14.0
8.5	14.8	8.8	13.7	9.7	15.1	9.5	15.2	10.3	14.9
10.9	13.2	7.6	13.7	9.6	15.8	8.8	13.0	11.2	16.4
9.9	15.7	9.2	15.2	9.6	14.5	8.2	13.6	9.6	12.5
11.3	13.8	7.6	12.0	10.1	12.4	9.0	14.3	9.4	14.1
12.0	14.9	8.8	11.7	7.5	12.2	10.7	13.9	11.3	16.0
9.6	14.2	9.3	14.9	10.1	13.7	9.7	13.0	12.0	15.0
8.3	12.6	8.9	13.8	8.7	13.3	9.2	15.9	8.5	12.3
9.2	12.4	10.6	15.2	9.9	12.8	7.0	12.3	10.1	13.1
8.4	14.7	9.8	12.6	9.7	13.8	8.0	12.1	10.6	16.6
11.7	16.1	9.7	10.9	10.4	16.0	8.5	14.3	11.4	14.3
10.0	12.2	9.5	14.8	8.9	11.8	11.4	14.6	9.6	15.3
8.2	12.6	10.8	13.5	9.5	11.9	9.3	14.8	11.1	13.4
11.3	14.8	9.7	11.5	10.5	16.8	9.9	13.1	10.4	15.1

TABLE 3.21   Delta=5 S.D., Variance Ratio=1, Correlation=0.9

X	Y	X	Y	X	Y	X	Y	X	Y
8.6	13.7	11.6	15.1	9.2	13.2	9.5	15.1	9.0	14.6
10.8	15.1	10.0	14.4	9.9	15.6	10.4	15.0	9.7	15.3
10.5	15.2	9.2	14.2	9.8	13.9	9.4	13.4	8.5	13.9
10.9	15.6	11.5	16.5	10.6	15.2	7.9	13.1	10.6	16.4
9.2	12.8	9.5	14.3	11.5	17.4	11.0	14.7	8.9	14.4
8.9	13.9	9.3	14.7	9.2	14.0	11.2	15.5	11.2	15.1
8.3	13.2	9.3	13.2	9.2	13.6	9.4	14.3	12.7	17.5
9.9	15.5	10.3	13.9	8.6	13.7	10.3	14.2	9.8	14.0
9.6	14.8	9.0	15.2	9.4	15.1	12.1	16.7	9.1	14.0
9.5	13.8	9.9	14.2	11.6	17.4	10.2	15.7	12.0	16.8
9.0	14.0	10.2	15.7	10.6	15.4	7.5	13.6	12.0	17.1
7.5	13.4	9.9	15.0	9.9	15.8	10.6	15.8	9.8	14.3
9.5	15.3	10.8	15.5	10.1	15.4	9.6	14.2	9.8	14.4
9.5	13.7	9.8	14.5	11.5	16.1	8.7	14.0	8.9	14.3
10.2	16.9	9.8	15.1	11.5	16.4	9.7	15.1	11.9	16.5
11.4	15.7	9.0	13.1	8.6	14.1	9.5	15.2	11.4	15.1
9.6	15.2	10.7	15.1	10.4	16.1	9.2	14.0	13.1	16.8
8.8	13.8	10.4	15.5	9.8	15.2	9.9	14.5	10.6	15.2
9.3	14.9	7.0	11.9	9.4	14.0	9.6	14.9	10.0	15.1
11.8	15.9	8.6	13.9	8.5	13.2	9.0	13.5	9.4	14.9
10.6	15.5	10.6	16.1	9.4	15.4	11.6	15.5	10.3	14.3
10.8	15.4	10.6	15.0	7.8	13.4	9.0	14.4	9.7	15.0
10.5	15.4	9.4	14.8	9.2	15.2	10.0	14.6	10.7	15.8
9.4	14.7	10.1	16.2	11.2	16.7	9.8	14.7	10.6	16.0
9.3	15.3	9.8	14.0	9.1	13.8	10.2	14.6	9.6	15.3

TABLE 3.22 Delta=5 S.D., Variance Ratio=2, Correlation=0

X	Y	X	Y	X	Y	X	Y	X	Y
8.3	16.5	8.2	14.4	10.9	14.7	14.2	15.3	9.9	11.5
9.0	13.2	8.3	14.6	10.3	13.9	9.6	16.7	10.8	14.9
13.2	16.7	8.7	14.7	7.4	11.7	7.1	15.2	10.0	17.8
8.1	12.3	9.8	16.5	11.0	16.7	11.5	14.7	10.7	18.8
8.5	12.7	12.7	17.2	9.9	12.0	12.8	14.3	11.3	13.2
9.4	14.3	8.7	13.9	10.7	15.4	12.4	14.3	8.9	16.5
11.3	17.5	7.3	13.8	11.0	14.4	11.9	15.1	11.3	15.3
10.7	15.6	7.2	16.6	9.3	17.4	11.8	16.7	11.2	14.9
11.1	17.4	12.7	14.3	10.1	15.3	10.4	13.6	10.3	14.9
10.9	14.9	9.9	14.9	8.9	14.3	10.7	14.3	10.8	17.6
9.6	13.4	9.4	14.0	9.1	12.9	9.7	14.9	11.5	16.3
11.0	13.3	12.4	17.1	9.9	11.7	12.3	18.8	8.5	13.4
9.4	12.8	10.1	14.6	9.8	18.7	9.8	14.8	10.1	11.3
11.5	16.3	8.5	17.1	8.7	12.1	9.8	15.3	10.9	19.4
10.6	18.6	10.4	16.4	9.6	15.3	6.3	14.2	9.9	17.9
10.2	12.2	10.2	14.3	9.8	10.4	8.7	15.1	10.9	17.3
11.1	15.5	12.4	15.6	9.4	13.1	9.8	18.1	8.7	16.5
10.9	15.5	9.5	10.6	7.4	14.6	12.2	14.3	10.1	16.6
8.3	13.0	9.2	13.9	9.4	17.5	7.0	15.7	9.5	14.0
11.4	14.2	11.0	17.9	10.2	15.4	6.8	13.4	10.6	13.2
10.3	13.5	9.1	15.1	9.6	15.2	10.6	16.1	9.7	13.2
10.4	11.8	10.5	14.3	12.5	11.7	11.1	15.0	9.4	14.6
9.0	16.8	10.7	13.2	10.9	14.9	8.1	13.6	10.4	11.3
9.4	12.5	11.6	14.9	9.7	10.4	8.2	16.4	11.6	14.2
8.7	15.9	11.8	17.7	10.3	15.9	10.1	13.8	9.6	10.8

TABLE 3.23 Delta=5 S.D., Variance Ratio=2, Correlation=0.5

X	Y	X	Y	X	Y	X	Y	X	Y
9.2	14.1	8.8	14.3	9.5	14.0	11.0	15.1	11.1	15.2
12.7	14.9	9.8	14.2	8.6	13.8	11.2	16.8	9.8	16.5
12.0	14.9	7.5	15.5	8.5	13.6	9.9	14.9	8.8	14.3
10.6	15.1	10.1	13.4	10.9	14.3	11.6	15.6	8.4	16.3
9.0	15.1	11.7	12.9	11.1	15.2	9.1	16.2	10.0	14.8
9.0	15.3	11.8	15.7	8.9	13.1	10.4	14.7	9.7	16.3
11.0	14.3	11.3	16.1	10.1	15.0	10.2	16.6	9.8	15.3
8.6	13.3	11.3	15.0	10.0	14.8	10.3	14.6	9.6	12.7
8.7	18.9	8.1	13.5	9.3	15.2	10.3	14.4	9.6	13.0
10.7	16.1	10.7	15.8	11.2	12.5	13.3	15.9	8.7	14.1
10.4	14.9	11.3	16.3	10.2	14.9	10.2	12.7	9.8	16.0
10.1	15.6	10.6	16.6	8.2	13.7	11.6	14.1	8.8	15.6
10.2	13.9	11.0	16.7	11.0	17.3	10.7	18.2	9.0	12.2
9.2	14.5	10.7	14.6	11.3	11.7	6.5	15.3	10.8	17.5
8.6	15.5	10.4	15.1	8.4	16.3	8.6	16.3	8.9	13.4
8.8	15.3	9.7	14.8	10.3	16.3	10.5	15.9	10.7	15.2
9.4	15.4	11.1	16.9	9.6	13.1	9.1	16.0	11.8	14.5
10.0	15.2	13.4	16.3	10.1	16.7	10.6	14.4	10.0	16.4
9.4	14.1	10.7	16.8	9.2	18.7	11.6	19.1	9.8	17.4
7.5	12.8	12.4	13.6	9.4	13.8	10.3	16.9	10.3	16.2
11.9	14.5	8.1	15.9	9.8	13.5	11.3	17.2	10.8	17.7
10.1	14.8	10.6	17.5	10.1	17.9	8 5	10.0	10.2	12.3
9.5	14.9	9.6	19.9	11.1	15.4	9.5	14.9	9.8	17.1
9.7	13.4	10.3	16.8	9.6	14.9	10.5	13.3	10.4	19.0
11.1	14.4	8.7	14.5	9.2	14.8	8.9	14.3	8.3	16.5

TABLE 3.24   Delta=5 S.D., Variance Ratio=2, Correlation=0.9

X	Y	X	Y	X	Y	X	Y	X	Y
9.6	13.7	10.3	15.4	10.1	15.6	10.5	16.4	10.9	14.6
11.0	15.1	9.4	15.5	9.2	14.6	10.0	13.5	9.7	15.4
8.7	12.6	9.1	13.4	9.9	13.6	9.0	14.3	10.6	15.6
11.8	16.8	9.1	13.9	9.6	15.2	9.3	13.8	9.8	14.8
10.0	14.6	11.4	16.2	9.8	15.0	10.4	15.2	8.4	13.5
8.8	13.4	9.8	14.5	10.2	15.2	9.8	15.3	11.7	16.5
10.0	13.9	10.4	16.8	9.4	14.6	10.1	16.5	10.2	13.7
8.9	14.6	11.0	16.3	9.9	14.3	12.1	16.9	10.2	14.8
9.2	13.6	10.4	15.3	12.6	16.4	10.8	15.6	9.4	14.7
9.4	14.2	9.5	12.4	10.6	14.7	10.2	14.3	10.4	15.6
11.5	16.5	10.5	15.4	10.1	15.3	9.4	15.8	10.2	16.4
10.6	14.7	10.6	15.8	9.3	15.0	12.2	17.1	10.4	17.1
9.6	14.5	7.0	11.1	8.2	11.8	10.0	14.7	11.0	16.5
8.5	14.1	8.7	12.2	12.2	17.4	8.9	13.3	10.7	15.6
11.0	15.4	12.1	16.7	9.6	13.5	10.6	14.4	9.5	15.5
9.8	14.8	10.0	15.4	7.4	12.5	10.2	16.1	10.0	16.0
8.0	13.7	11.6	17.3	9.6	14.4	9.6	14.0	8.9	12.3
9.5	13.8	10.7	15.9	7.9	12.5	10.5	15.5	10.7	16.3
9.0	14.8	9.9	14.6	8.9	13.6	8.1	11.8	11.1	17.2
9.7	13.7	9.6	14.1	9.2	11.9	11.2	16.2	9.8	15.4
11.3	16.3	10.1	15.0	9.6	14.3	10.2	16.1	11.3	15.8
10.3	14.5	10.7	16.2	10.5	15.3	12.4	18.0	9.9	13.6
11.3	16.4	11.0	16.1	9.6	16.0	8.4	13.1	9.2	14.4
11.5	17.6	10.0	14.8	9.9	15.0	9.1	12.5	12.6	18.4
9.9	16.9	9.8	14.6	11.0	16.2	11.2	15.4	10.0	15.6

TABLE 3.25   Delta=5 S.D., Variance Ratio=5, Correlation=0

X	Y	X	Y	X	Y	X	Y	X	Y
8.1	17.3	9.1	12.4	6.4	15.8	9.3	10.8	11.1	18.2
9.6	18.3	11.6	10.5	10.0	10.3	10.9	8.7	9.0	15.2
10.6	19.7	8.6	21.5	12.1	16.1	10.5	16.6	8.4	16.0
9.4	22.1	12.1	19.4	7.6	19.2	11.7	17.1	11.5	18.6
9.5	11.5	10.4	19.2	13.2	15.3	10.3	19.1	11.2	15.5
11.6	12.7	12.5	12.9	12.0	13.6	11.5	11.1	10.9	13.6
10.9	15.3	9.2	16.4	9.5	19.3	11.4	5.7	10.3	18.1
10.9	17.0	10.0	10.0	7.9	15.1	10.3	14.8	9.3	16.5
7.5	16.5	10.1	15.1	12.3	14.3	10.6	15.1	9.5	13.4
11.8	14.5	10.0	11.8	9.9	10.2	11.8	11.4	10.1	8.6
10.6	13.8	6.9	18.0	10.3	13.0	11.3	15.1	7.6	14.5
9.3	18.9	11.2	21.0	9.5	19.2	10.5	13.5	10.6	11.7
8.2	15.2	10.2	15.7	7.2	19.7	10.1	14.3	12.1	13.8
6.2	15.0	10.5	12.0	10.9	15.5	8.8	13.2	9.9	14.9
8.8	16.2	9.3	16.9	9.5	14.3	11.2	16.6	9.6	15.2
10.3	8.9	8.1	15.3	5.0	13.6	10.3	9.0	9.8	15.8
10.3	9.8	9.4	8.8	9.2	17.7	9.5	14.2	9.3	16.7
9.3	17.3	8.9	8.6	8.5	10.9	10.2	14.6	12.2	15.9
11.7	14.7	11.4	9.6	12.2	7.9	8.9	11.8	8.9	15.2
12.2	7.0	11.1	12.8	9.3	15.4	10.3	15.1	9.3	14.1
11.3	16.1	11.5	12.0	11.5	16.8	10.8	15.4	11.1	11.6
14.1	15.8	9.2	17.6	8.4	11.9	10.0	21.1	8.4	9.3
9.1	16.0	9.2	12.4	11.8	19.5	11.6	10.2	11.7	14.7
11.4	14.8	12.4	6.2	8.7	8.9	10.1	12.7	11.0	16.8
10.3	14.3	10.0	14.2	9.8	15.9	9.4	11.4	9.3	15.3

TABLE 3.26 Delta=5 S.D., Variance Ratio=5, Correlation=0.5

X	Y	X	Y	X	Y	X	Y	X	Y
9.9	16.6	11.1	22.4	10.1	16.8	10.5	16.3	10.3	18.7
10.8	17.5	10.1	17.5	9.4	15.1	12.2	16.7	8.5	13.2
11.3	10.4	10.0	13.1	10.7	18.9	10.4	14.3	11.6	20.8
10.0	17.9	10.0	16.5	10.5	12.6	9.7	8.1	10.7	12.1
10.4	14.3	9.3	14.9	9.0	11.7	9.4	13.7	9.1	16.5
12.1	19.1	10.0	18.8	11.2	16.0	10.6	14.4	7.5	13.0
10.6	16.4	10.8	14.1	11.6	15.8	10.1	9.0	11.2	17.5
9.9	11.3	10.8	16.4	7.9	11.6	9.4	14.0	8.3	11.7
10.8	20.1	9.5	15.8	9.0	9.7	8.6	12.8	10.9	19.8
7.2	9.6	8.8	15.4	8.9	14.5	10.3	5.7	10.1	19.1
9.3	15.9	11.7	16.3	8.1	13.4	11.8	16.2	10.4	15.5
9.4	13.3	9.5	15.2	10.4	13.3	10.3	10.1	9.9	15.6
10.2	13.1	9.9	13.8	10.7	16.8	9.9	16.3	9.0	17.5
12.8	16.6	11.4	22.4	7.8	12.5	8.4	9.6	9.1	15.7
9.2	15.4	9.5	14.7	10.9	17.0	10.6	14.8	9.6	14.3
8.8	13.5	11.8	16.3	9.9	14.4	8.1	14.4	10.2	13.2
10.8	14.0	8.6	7.9	8.3	13.3	10.1	18.3	11.8	17.5
8.0	19.1	10.7	9.9	9.8	13.7	12.0	15.3	9.4	14.4
9.0	12.0	9.7	17.4	9.5	11.7	9.4	17.4	10.2	17.5
9.6	15.2	11.4	13.0	10.8	21.8	8.7	15.0	11.0	15.8
12.1	20.4	10.4	15.6	9.1	13.5	8.3	14.6	8.1	12.7
11.3	10.2	9.7	15.5	9.1	15.5	9.4	13.5	10.7	16.9
9.5	16.0	10.3	12.7	9.5	11.3	9.0	20.9	12.3	12.3
10.2	18.6	7.1	9.9	11.4	15.7	11.2	18.8	9.6	14.5
9.5	17.3	10.9	13.7	8.5	12.1	10.0	14.4	10.5	13.6

TABLE 3.27 Delta=5 S.D., Variance Ratio=5, Correlation=0.9

X	Y	X	Y	X	Y	X	Y	X	Y
10.4	13.7	12.0	19.6	8.8	13.3	10.1	16.6	9.1	14.4
8.7	12.5	11.3	18.4	10.4	15.1	9.4	14.8	11.7	17.1
10.5	13.5	10.8	15.8	10.8	15.3	10.6	17.8	9.2	11.4
10.0	17.7	10.7	15.7	9.5	13.7	9.7	11.0	11.3	15.8
10.1	15.0	10.8	16.9	11.2	15.4	9.9	14.8	10.9	17.8
8.4	11.9	12.0	18.7	8.3	11.4	7.3	10.6	10.2	15.8
9.7	14.2	7.8	9.8	9.9	11.9	12.8	21.7	10.6	15.5
9.8	14.5	8.4	11.8	11.8	17.5	10.3	16.2	8.5	13.6
9.9	15.1	10.3	16.6	7.7	8.3	9.7	14.2	8.2	11.3
8.6	11.6	12.5	20.5	10.7	17.0	11.2	17.2	9.3	14.3
11.1	18.0	11.8	19.1	10.0	14.7	10.0	13.9	10.2	14.5
8.6	11.8	9.8	12.3	11.3	17.0	10.0	15.2	9.5	13.7
10.8	16.6	10.1	14.5	8.6	12.2	8.6	13.0	10.6	17.2
10.5	16.0	10.8	15.0	11.3	16.1	7.7	11.3	9.9	13.9
9.5	14.7	10.4	16.9	11.4	18.4	9.5	15.0	10.7	14.8
10.2	17.7	10.1	14.9	10.2	14.9	10.8	15.2	11.4	21.0
9.7	13.0	9.5	13.9	9.5	15.2	9.9	15.2	11.0	16.3
11.2	15.3	9.3	14.5	9.5	13.7	10.7	13.6	9.4	12.8
9.1	13.0	10.8	18.1	8.3	9.5	9.3	16.3	7.5	9.5
11.3	16.7	9.5	15.0	10.4	15.8	9.2	13.8	11.6	18.3
9.1	15.9	10.2	14.2	10.6	16.8	7.7	10.9	9.9	15.4
9.3	12.5	10.6	15.5	10.5	16.9	10.5	17.2	9.0	13.2
8.1	11.6	10.9	16.1	8.4	9.7	10.4	15.3	9.7	15.0
10.5	13.5	9.7	14.3	10.5	14.5	9.3	11.8	11.1	15.6
10.6	16.4	9.0	13.3	9.9	13.7	11.0	17.6	9.9	13.3

EXERCISES

Exercise 3.1:   Comparison of Means

Starting at random in the table, pick a sequence of 50 pairs of
numbers ("wrapping around" to the beginning if necessary). Run the
following on the first 10 pairs, then on the first 30 pairs, and
finally on all 50 pairs:

    Test the hypothesis of equal mean with a sign test, then with
a Wilcoxon Signed Rank Test, and, finally, with a paired t-test.
The student should go through this sequence (sign test to Wilcoxon
to paired t-) and try to determine when the difference in mean or
the degree of correlation is so great that the sign test is adequate
to reach a firm conclusion. The student should also watch how the
variance ratio can affect the relative sensitivity (or power) of the
Wilcoxon and paired t-tests.

Exercise 3.2:   Regression

Starting at random in the table, pick a sequence of 50 number pairs
("wrapping around" to the beginning if necessary). Run the following on the first 10 pairs, then on the first 30 pairs, and, finally,
on all 50 pairs:

    Order the pairs of data in terms of increasing X. Plot the
points on rectangular graph paper and calculate the Least Squares
estimators of the linear function

    $Y = A + BX$

    Plot the Least Squares line on the graph paper: divide the X-
values at their median and locate the median value of Y for the
smaller values of X and the median value of Y for the larger values
of X. Connect the two points thus defined. Compare this line to
the Least Squares line.

    Examine the scatter of data points about the Least Squares line
and also about the line connecting the two medians. Plot parallel
lines one or two standard deviations (using residual variance off

the line) on either side of the Least Squares line. Note how many of the points fall within and without those bounds. This will provide the student with some sense of the way in which individual points tend to scatter away from the "best fitting line."

Choose a second random sample of the same size as the one used in the computation of the Least Squares line. Plot the points from the second random sample, but with a different symbol, on the same sheet of graph paper. Compare the scatter of this second sample about the Least Squares line from the first sample. Note how good is the prediction of one sample from the other and how the quality of that prediction changes with different correlations and variance ratios. What is the effect of sample size on the scatter? on the consistency of the regression line from one sample to the other?

STATISTICAL TECHNIQUES AND REFERENCES
Sign Test

This is a one sample test of the hypothesis that the median is equal to some fixed predetermined value. If the sample consists of paired differences, then the assumed median is 0.0. Count all values that are less than or equal to the assumed median. That count Z should be distributed as a binomial with $p = 1/2$.

References

Dixon and Massey, Section 17-1, pp. 335-339.
Hollander and Wolfe, Section 3.4, pp. 39-49.
Kendall and Stuart, Sections 32.2-32.10, pp. 513-518.
Miller and Freund, Section 10.2, pp. 272-275.
Owen, Section 12.1, pp. 362-366.

LEAST SQUARES ESTIMATION OF LINEAR REGRESSION

The model here is

$$E(Y:X) = A + BX$$

References

Dixon and Massey, Chapter 11, pp. 193-200.
Mood and Graybill, Chapter 13, pp. 328-342.
Miller and Friend, Chapter 11, pp. 289-305.
Raktoe and Hubert, Chapter 14, pp. 269-283.

SELECTED WORKED-OUT EXAMPLES

Exercise 3.1 applied to Table 3.3.

Starting in column 3, row 5:

	$n = 10$	$n = 30$	$n = 50$
Sign tests:			
# times X greater than Y/			
# times $X \neq Y$	7/10	15/27	26/45
one-sided sig	0.17	0.35	0.19
paired t tests:			
mean X - mean Y	0.130	0.040	0.042
Var(paired deltas)	0.231	0.478	0.505
paired t tests	0.85	0.32	0.35
Wilcoxon Signed Rank Tests:			
Signed Rank Sum = W	38.0	201.0	509.5
$E(W)$	27.5	232.5	637.5
$z=(W-E(W))$/SD	1.07	- 0.65	- 1.24

Note how these three tests all agree qualitatively (in terms of
degree of statistical significance) although they address slightly
different questions.

Exercise 3.1 applied to Table 3.9.

   Starting in Column 2, row 10:

	n = 10	n = 30	n = 50
sign tests:			
# times X greater than Y/			
# times X≠Y	7/10	16/30	26/50
one-sided sig.	0.17	0.43	0.44
paired t tests:			
mean X - mean Y	0.680	- 0.257	- 0.144
Var(paired delta)	4.186	4.073	3.189
paired t tests	1.05	- 0.70	- 0.57
Wilcoxon Signed Rank Tests:			
Signed Rank Sum = W	37.5	207.0	601.0
E(W)	27.5	232.5	637.5
z=(W-E)W))/SD	1.02	- 0.52	- 0.35

Note how the increase in variance from Table 3.3 affects the denominator of the paired *t* test and reduces the number of zeroes (sign test) and ties (Wilcoxon test).

Exercise 3.1 (restricted to first 10 values) for Tables 3.1, 3.10, 3.16:

Table	3.1	3.10	3.16
Starting col, row	3,8	3,7	1,5

sign tests:
# times X greater than Y/			
# times $X \neq Y$	2/10	6/10	5/10

paired t tests:
mean X - mean Y	0.580	- 1.710	0.250
Var(paired delta)	2.195	4.217	21.076
paired t	1.24	- 2.63	0.17

Wilcoxon Tests:
Signed Rank Sum = W	44.0	10.0	25.0
$z=(W-E(W))/SD$	1.68	- 1.78	- 0.25

values of $X(i)-Y(i)$:	1.5	- 1.7	- 2.8
	1.1	0.3	1.9
	3.3	0.1	- 5.8
	1.0	- 1.9	8.4
	- 2.6	0.4	- 3.5
	- 0.3	- 3.8	6.6
	0.3	- 2.9	1.2
	0.5	0.7	- 2.4
	0.3	- 3.0	- 2.9

4

CONTINGENCY TABLES

INTRODUCTION

Although statistical distributions deal with the scatter of numbers,
the statistician is often called upon to analyze non-numerical patterns in the classification of events. These reduce to numbers by
organizing counts of the events into contingency tables. In its
most general form, we observe a collection of "individuals," each of
whom can be characterized into one of a fixed (and often large) number of cells. We count the individuals in each cell and arrange
this pattern of counts in an organized fashion. It may be that some
of the cells can be thought of as subsets of a larger class. For
instance, male and female are subsets of the class of sexes. Two or
more of these classes may also be cross-indexed. For instance, the
males and females may also be organized into job classifications.
The entire set of individuals can then be displayed as an array or
contingency table, like this:

<table>
<tr><td></td><td colspan="4">Job Classification</td></tr>
<tr><td></td><td colspan="2">Nonsupervisory</td><td colspan="2">Supervisory</td></tr>
<tr><td></td><td>Blue Collar</td><td>White Collar</td><td>Blue Collar</td><td>White Collar</td></tr>
<tr><td>Male</td><td>463</td><td>95</td><td>44</td><td>59</td></tr>
<tr><td>Female</td><td>12</td><td>214</td><td>6</td><td>25</td></tr>
</table>

Although it is possible to find other classes and cross tabuluua-
tions that will yield more complex dimensions to the table, the individuals who derived or counted the original data may be wedded to a

specific scheme of classification. Sometimes, the scheme of classification is the result of theoretical knowledge about the underlying
scientific problem. Far more often, the scheme of classification is
purely *ad hoc* and dependent more on convenience and the lack of
imagination of the collector than on anything else.

   A good statistician should come to the data with an ability to
generalize beyond the classifications presented. To the statistician, the contingency table is nothing but a collection of counts.
A specific hypothesis to be examined will suggest a structure for
those counts, but the essential information in a contingency table
is sometimes ignored in analysis because the hypotheses examined
were based upon the arbitrary classifications that were imposed on
the counts for convenience of tabulation or display. The statistician should be alert to spot other internal structures in the
tables that might be ignored by "standard" statistical tests if they
were applied blindly to the structure presented.

   Thus, a good statistician must be skeptical about the form in
which a contingency table is presented. Are the classifications
chosen for display appropriate to the questions being asked? Even
when the display is appropriate, there may be an internal structure
in the display that may not be obvious to a standard statistical
test, like the chi square test for independence of rows and columns.

   However, how does the statistician know when the apparent
structure within a table is indicative of real structure and when it
is merely random noise? This chapter will provide some slight experience in that direction.

   There are a large number of possible structures for contingency
tables and there is no way of providing experience for even a
"representative" set of them. Instead, we will be content here to
examine a particular one, in the belief that the experience gained
this way will aid the student in the examination of larger and more
complicated tables. All the data in Tables 4.1-4.12 are arranged in
the form of a 2 x 4 contingency table as follows:

	Categories			
	A	B	C	D
Yes	3	1	2	4
No	9	10	11	8

For Tables 4.1, 4.5, and 4.9, the probability of "yes" is constant across all four categories (and the probability of "no" is its complement). For Tables 4.2, 4.6, and 4.10, there is a slight linear increase in probability of "yes" from Category A to B to C to D. For Tables 4.3, 4.7, and 4.11, the increase (called a "dose response" in the tables) is moderate. For Tables 4.4, 4.8, and 4.12, the "dose response" is extreme.

Most theoretical discussions of contingency tables assume that the counts of events follow a multinomial distribution, that is, that all individuals have the same probability of falling into a given cell. This means that the only random components in the table consist of events that occur with the same probability for all individuals in a given class. In practice, this constant probability assumption seldom exists. Instead, the individuals in a class often represent a sample with slightly differing probabilities of event. For instance, patients might be assigned to four treatments (our categories A, B, C, and D) and the successful outcome of a treatment ("yes" in our tables) is noted. Because of differing baseline characteristics, no two patients have the same probability of success for a given treatment, so there are several probabilities of "yes" associated with any specific category.

As a result, we are really looking at mixtures of mutinomials, rather than pure multinomial distributions. It can be shown that the expectation of response is the same in both the mixed and pure models. However, the variance of response is smaller in the mixed model. An extreme form of this would occur if half the patients entering into treatment A had no chance of success and the other half were sure of success. We would have a mixture of two binomials, one with $p=0$ and the other with $p=1$. The variance of the resulting count would be zero (since all results were predetermined), yet the expectation would be 1/2 for both this mixed and a similar pure binomial model.

To illustrate what this variance reduction does to data. Tables 4.5-4.8 have a slight mixture of binomials for each category, and Tables 4.9-4.12 have a moderate degree of mixing. The student should be alert to the degree to which this mixing affects statistical procedures.

58

TABLE 4. 1 — PURE BINOMIAL COMPARISONS — NO DOSE RESPONSE

	CATEGORIES					CATEGORIES					CATEGORIES		
	A	B	C	D		A	B	C	D		A	B	C	D
YES	8	3	3	2		4	1	1	4		4	4	6	4
NO	4	9	9	10		8	11	11	8		8	8	6	8

	CATEGORIES					CATEGORIES					CATEGORIES		
	A	B	C	D		A	B	C	D		A	B	C	D
YES	4	6	4	2		3	3	5	2		5	3	2	5
NO	8	6	8	10		9	9	7	10		7	9	10	7

	CATEGORIES					CATEGORIES					CATEGORIES		
	A	B	C	D		A	B	C	D		A	B	C	D
YES	5	4	3	6		6	4	2	4		4	2	4	7
NO	7	8	9	6		6	8	10	8		8	10	8	5

	CATEGORIES					CATEGORIES					CATEGORIES		
	A	B	C	D		A	B	C	D		A	B	C	D
YES	4	4	6	4		2	5	1	1		3	2	3	2
NO	8	8	6	8		10	7	11	11		9	10	9	10

	CATEGORIES					CATEGORIES					CATEGORIES		
	A	B	C	D		A	B	C	D		A	B	C	D
YES	2	3	2	2		6	5	3	5		2	1	4	5
NO	10	9	10	10		6	7	9	7		10	11	8	7

	CATEGORIES					CATEGORIES					CATEGORIES		
	A	B	C	D		A	B	C	D		A	B	C	D
YES	4	1	4	5		3	3	3	5		6	1	3	3
NO	8	11	8	7		9	9	9	7		6	11	9	9

	CATEGORIES					CATEGORIES					CATEGORIES		
	A	B	C	D		A	B	C	D		A	B	C	D
YES	4	4	5	4		7	2	4	5		4	4	5	3
NO	8	8	7	8		5	10	8	7		8	8	7	9

	CATEGORIES					CATEGORIES					CATEGORIES		
	A	B	C	D		A	B	C	D		A	B	C	D
YES	1	5	2	1		5	3	2	2		4	2	6	4
NO	11	7	10	11		7	9	10	10		8	10	6	8

TABLE 4. 2   PURE BINOMIAL COMPARISONS   SLIGHT DOSE RESPONSE

	CATEGORIES					CATEGORIES					CATEGORIES		
	A	B	C	D	A	B	C	D	A	B	C	D
YES	3	3	4	5	5	7	6	8	3	4	4	4
NO	9	9	8	7	7	5	6	4	9	8	8	8

	CATEGORIES					CATEGORIES					CATEGORIES		
	A	B	C	D	A	B	C	D	A	B	C	D
YES	4	4	4	3	3	3	5	7	2	3	5	5
NO	8	8	8	9	9	9	7	5	10	9	7	7

	CATEGORIES					CATEGORIES					CATEGORIES		
	A	B	C	D	A	B	C	D	A	B	C	D
YES	5	7	6	7	3	2	4	8	4	3	5	4
NO	7	5	6	5	9	10	8	4	8	9	7	8

	CATEGORIES					CATEGORIES					CATEGORIES		
	A	B	C	D	A	B	C	D	A	B	C	D
YES	4	5	4	4	3	5	2	6	4	3	6	3
NO	8	7	8	8	9	7	10	6	8	9	6	9

	CATEGORIES					CATEGORIES					CATEGORIES		
	A	B	C	D	A	B	C	D	A	B	C	D
YES	3	5	7	5	2	2	4	6	1	8	6	7
NO	9	7	5	7	10	10	8	6	11	4	6	5

	CATEGORIES					CATEGORIES					CATEGORIES		
	A	B	C	D	A	B	C	D	A	B	C	D
YES	6	5	6	6	4	3	4	6	5	4	5	2
NO	6	7	6	6	8	9	8	6	7	8	7	10

	CATEGORIES					CATEGORIES					CATEGORIES		
	A	B	C	D	A	B	C	D	A	B	C	D
YES	3	3	1	5	8	3	6	7	3	6	3	7
NO	9	9	11	7	4	9	6	5	9	6	9	5

	CATEGORIES					CATEGORIES					CATEGORIES		
	A	B	C	D	A	B	C	D	A	B	C	D
YES	1	6	6	6	4	4	4	6	1	5	5	6
NO	11	6	6	6	8	8	8	6	11	7	7	6

TABLE 4. 3   PURE BINOMIAL COMPARISONS MODERATE DOSE RESPONSE

	CATEGORIES				CATEGORIES				CATEGORIES			
	A	B	C	D	A	B	C	D	A	B	C	D
YES	3	5	5	5	6	4	4	6	6	7	5	4
NO	9	7	7	7	6	8	8	6	6	5	7	8

	CATEGORIES				CATEGORIES				CATEGORIES			
	A	B	C	D	A	B	C	D	A	B	C	D
YES	1	3	5	5	5	6	5	8	7	5	9	9
NO	11	9	7	7	7	6	7	4	5	7	3	3

	CATEGORIES				CATEGORIES				CATEGORIES			
	A	B	C	D	A	B	C	D	A	B	C	D
YES	1	4	4	9	5	9	5	5	3	6	6	7
NO	11	8	8	3	7	3	7	7	9	6	6	5

	CATEGORIES				CATEGORIES				CATEGORIES			
	A	B	C	D	A	B	C	D	A	B	C	D
YES	4	5	6	7	6	6	7	6	4	4	8	8
NO	8	7	6	5	6	6	5	6	8	8	4	4

	CATEGORIES				CATEGORIES				CATEGORIES			
	A	B	C	D	A	B	C	D	A	B	C	D
YES	2	5	5	6	3	4	3	4	3	2	5	5
NO	10	7	7	6	9	8	9	8	9	10	7	7

	CATEGORIES				CATEGORIES				CATEGORIES			
	A	B	C	D	A	B	C	D	A	B	C	D
YES	6	5	8	7	4	5	4	7	6	5	1	7
NO	6	7	4	5	8	7	8	5	6	7	11	5

	CATEGORIES				CATEGORIES				CATEGORIES			
	A	B	C	D	A	B	C	D	A	B	C	D
YES	5	6	7	7	5	2	8	5	5	4	5	8
NO	7	6	5	5	7	10	4	7	7	8	7	4

	CATEGORIES				CATEGORIES				CATEGORIES			
	A	B	C	D	A	B	C	D	A	B	C	D
YES	3	6	7	5	2	6	7	9	4	3	8	7
NO	9	6	5	7	10	6	5	3	8	9	4	5

TABLE 4. 4   PURE BINOMIAL COMPARISONS   SEVERE DOSE RESPONSE

	CATEGORIES					CATEGORIES					CATEGORIES			
	A	B	C	D		A	B	C	D		A	B	C	D
YES	5	5	6	11		6	6	9	12		3	3	8	11
NO	7	7	6	1		6	6	3	0		9	9	4	1
	A	B	C	D		A	B	C	D		A	B	C	D
YES	4	9	10	12		4	7	7	9		3	4	6	12
NO	8	3	2	0		8	5	5	3		9	8	6	0
	A	B	C	D		A	B	C	D		A	B	C	D
YES	5	8	9	11		1	6	10	9		3	5	9	9
NO	7	4	3	1		11	6	2	3		9	7	3	3
	A	B	C	D		A	B	C	D		A	B	C	D
YES	1	7	8	11		1	6	8	10		4	6	7	10
NO	11	5	4	1		11	6	4	2		8	6	5	2
	A	B	C	D		A	B	C	D		A	B	C	D
YES	4	8	8	12		4	10	9	10		3	5	12	10
NO	8	4	4	0		8	2	3	2		9	7	0	2
	A	B	C	D		A	B	C	D		A	B	C	D
YES	5	6	8	10		6	5	8	11		5	8	10	11
NO	7	6	4	2		6	7	4	1		7	4	2	1
	A	B	C	D		A	B	C	D		A	B	C	D
YES	4	6	11	10		6	6	10	11		4	5	10	12
NO	8	6	1	2		6	6	2	1		8	7	2	0
	A	B	C	D		A	B	C	D		A	B	C	D
YES	1	8	9	11		4	6	5	8		1	4	9	10
NO	11	4	3	1		8	6	7	4		11	8	3	2

TABLE 4. 5   SLIGHT MIX BINOMIAL COMPARISONS   NO DOSE RESPONSE

	CATEGORIES					CATEGORIES					CATEGORIES			
	A	B	C	D		A	B	C	D		A	B	C	D
YES	4	7	4	2		2	4	3	1		3	3	3	6
NO	8	5	8	10		10	8	9	11		9	9	9	6

	CATEGORIES					CATEGORIES					CATEGORIES			
	A	B	C	D		A	B	C	D		A	B	C	D
YES	6	2	4	1		5	5	2	4		2	2	6	5
NO	6	10	8	11		7	7	10	8		10	10	6	7

	CATEGORIES					CATEGORIES					CATEGORIES			
	A	B	C	D		A	B	C	D		A	B	C	D
YES	5	3	3	2		3	5	2	4		2	4	3	5
NO	7	9	9	10		9	7	10	8		10	8	9	7

	CATEGORIES					CATEGORIES					CATEGORIES			
	A	B	C	D		A	B	C	D		A	B	C	D
YES	3	6	2	1		4	8	6	2		6	6	1	5
NO	9	6	10	11		8	4	6	10		6	6	11	7

	CATEGORIES					CATEGORIES					CATEGORIES			
	A	B	C	D		A	B	C	D		A	B	C	D
YES	4	3	2	4		2	4	6	5		4	3	5	7
NO	8	9	10	8		10	8	6	7		8	9	7	5

	CATEGORIES					CATEGORIES					CATEGORIES			
	A	B	C	D		A	B	C	D		A	B	C	D
YES	5	2	2	2		8	3	5	3		3	0	3	0
NO	7	10	10	10		4	9	7	9		9	12	9	12

	CATEGORIES					CATEGORIES					CATEGORIES			
	A	B	C	D		A	B	C	D		A	B	C	D
YES	3	4	2	3		5	5	5	4		5	5	2	4
NO	9	8	10	9		7	7	7	8		7	7	10	8

	CATEGORIES					CATEGORIES					CATEGORIES			
	A	B	C	D		A	B	C	D		A	B	C	D
YES	2	1	4	4		2	5	3	2		5	3	7	5
NO	10	11	8	8		10	7	9	10		7	9	5	7

TABLE 4. 6   SLIGHT MIX BINOMIAL COMPARISONS   SLIGHT DOSE RESPONSE

	CATEGORIES					CATEGORIES					CATEGORIES			
	A	B	C	D		A	B	C	D		A	B	C	D
YES	6	5	5	4		2	3	6	8		2	8	4	7
NO	6	7	7	8		10	9	6	4		10	4	8	5

	CATEGORIES					CATEGORIES					CATEGORIES			
	A	B	C	D		A	B	C	D		A	B	C	D
YES	2	4	3	3		4	4	3	7		4	4	5	4
NO	10	8	9	9		8	8	9	5		8	8	7	8

	CATEGORIES					CATEGORIES					CATEGORIES			
	A	B	C	D		A	B	C	D		A	B	C	D
YES	6	1	5	7		3	7	6	4		5	6	7	3
NO	6	11	7	5		9	5	6	8		7	6	5	9

	CATEGORIES					CATEGORIES					CATEGORIES			
	A	B	C	D		A	B	C	D		A	B	C	D
YES	3	4	5	7		3	2	7	4		5	4	5	6
NO	9	8	7	5		9	10	5	8		7	8	7	6

	CATEGORIES					CATEGORIES					CATEGORIES			
	A	B	C	D		A	B	C	D		A	B	C	D
YES	4	4	6	6		6	7	4	5		6	5	6	7
NO	8	8	6	6		6	5	8	7		6	7	6	5

	CATEGORIES					CATEGORIES					CATEGORIES			
	A	B	C	D		A	B	C	D		A	B	C	D
YES	3	1	3	6		2	2	6	4		6	2	3	8
NO	9	11	9	6		10	10	6	8		6	10	9	4

	CATEGORIES					CATEGORIES					CATEGORIES			
	A	B	C	D		A	B	C	D		A	B	C	D
YES	7	5	9	3		7	3	6	5		3	3	5	6
NO	5	7	3	9		5	9	6	7		9	9	7	6

	CATEGORIES					CATEGORIES					CATEGORIES			
	A	B	C	D		A	B	C	D		A	B	C	D
YES	3	1	5	6		2	5	4	8		4	6	4	3
NO	9	11	7	6		10	7	8	4		8	6	8	9

TABLE 4. 7 SLIGHT MIX BINOMIAL COMPARISONS MODERATE DOSE RESPONSE

	CATEGORIES				CATEGORIES				CATEGORIES			
	A	B	C	D	A	B	C	D	A	B	C	D
YES	3	4	7	6	2	2	3	8	4	5	5	10
NO	9	8	5	6	10	10	9	4	8	7	7	2

	CATEGORIES				CATEGORIES				CATEGORIES			
	A	B	C	D	A	B	C	D	A	B	C	D
YES	2	5	9	8	5	5	4	8	5	7	5	8
NO	10	7	3	4	7	7	8	4	7	5	7	4

	CATEGORIES				CATEGORIES				CATEGORIES			
	A	B	C	D	A	B	C	D	A	B	C	D
YES	8	5	6	10	3	5	4	5	4	7	6	5
NO	4	7	6	2	9	7	8	7	8	5	6	7

	CATEGORIES				CATEGORIES				CATEGORIES			
	A	B	C	D	A	B	C	D	A	B	C	D
YES	2	3	5	5	4	4	6	9	3	4	6	8
NO	10	9	7	7	8	8	6	3	9	8	6	4

	CATEGORIES				CATEGORIES				CATEGORIES			
	A	B	C	D	A	B	C	D	A	B	C	D
YES	2	5	6	9	5	2	8	5	9	3	6	6
NO	10	7	6	3	7	10	4	7	3	9	6	6

	CATEGORIES				CATEGORIES				CATEGORIES			
	A	B	C	D	A	B	C	D	A	B	C	D
YES	2	5	7	6	3	6	6	6	7	4	7	9
NO	10	7	5	6	9	6	6	6	5	8	5	3

	CATEGORIES				CATEGORIES				CATEGORIES			
	A	B	C	D	A	B	C	D	A	B	C	D
YES	1	5	6	9	2	7	8	9	3	4	8	7
NO	11	7	6	3	10	5	4	3	9	8	4	5

	CATEGORIES				CATEGORIES				CATEGORIES			
	A	B	C	D	A	B	C	D	A	B	C	D
YES	4	5	5	7	4	5	2	6	4	5	10	10
NO	8	7	7	5	8	7	10	6	8	7	2	2

TABLE 4. 8 SLIGHi MIX BINOMIAL COMPARISONS   SEVERE DOSE RESPONSE

	CATEGORIES					CATEGORIES					CATEGORIES			
	A	B	C	D		A	B	C	D		A	B	C	D
YES	6	4	10	10		4	7	6	12		6	6	8	11
NO	6	8	2	2		8	5	6	0		6	6	4	1

	CATEGORIES					CATEGORIES					CATEGORIES			
	A	B	C	D		A	B	C	D		A	B	C	D
YES	2	6	7	11		4	5	7	11		4	5	8	11
NO	10	6	5	1		8	7	5	1		8	7	4	1

	CATEGORIES					CATEGORIES					CATEGORIES			
	A	B	C	D		A	B	C	D		A	B	C	D
YES	2	6	6	10		4	8	10	11		3	3	6	11
NO	10	6	6	2		8	4	2	1		9	9	6	1

	CATEGORIES					CATEGORIES					CATEGORIES			
	A	B	C	D		A	B	C	D		A	B	C	D
YES	4	7	7	11		1	5	11	8		5	6	5	11
NO	8	5	5	1		11	7	1	4		7	6	7	1

	CATEGORIES					CATEGORIES					CATEGORIES			
	A	B	C	D		A	B	C	D		A	B	C	D
YES	2	2	10	11		3	6	8	11		3	8	10	12
NO	10	10	2	1		9	6	4	1		9	4	2	0

	CATEGORIES					CATEGORIES					CATEGORIES			
	A	B	C	D		A	B	C	D		A	B	C	D
YES	1	7	11	10		5	8	8	9		4	5	6	10
NO	11	5	1	2		7	4	4	3		8	7	6	2

	CATEGORIES					CATEGORIES					CATEGORIES			
	A	B	C	D		A	B	C	D		A	B	C	D
YES	4	3	10	11		6	3	11	9		3	6	10	11
NO	8	9	2	1		6	9	1	3		9	6	2	1

	CATEGORIES					CATEGORIES					CATEGORIES			
	A	B	C	D		A	B	C	D		A	B	C	D
YES	3	5	9	10		4	3	8	11		3	7	8	9
NO	9	7	3	2		8	9	4	1		9	5	4	3

TABLE 4. 9   MODER. MIX BINOMIAL COMPARISONS   NO DOSE RESPONSE

	CATEGORIES					CATEGORIES					CATEGORIES			
	A	B	C	D		A	B	C	D		A	B	C	D
YES	3	3	6	5		6	6	4	4		4	0	4	5
NO	9	9	6	7		6	6	8	8		8	12	8	7

	CATEGORIES					CATEGORIES					CATEGORIES			
	A	B	C	D		A	B	C	D		A	B	C	D
YES	2	5	5	5		2	3	4	3		4	4	3	2
NO	10	7	7	7		10	9	8	9		8	8	9	10

	CATEGORIES					CATEGORIES					CATEGORIES			
	A	B	C	D		A	B	C	D		A	B	C	D
YES	5	5	4	6		5	5	2	4		1	4	4	4
NO	7	7	8	6		7	7	10	8		11	8	8	8

	CATEGORIES					CATEGORIES					CATEGORIES			
	A	B	C	D		A	B	C	D		A	B	C	D
YES	8	2	6	4		6	3	6	6		4	2	5	3
NO	4	10	6	8		6	9	6	6		8	10	7	9

	CATEGORIES					CATEGORIES					CATEGORIES			
	A	B	C	D		A	B	C	D		A	B	C	D
YES	4	4	0	1		3	1	7	3		3	6	4	3
NO	8	8	12	11		9	11	5	9		9	6	8	9

	CATEGORIES					CATEGORIES					CATEGORIES			
	A	B	C	D		A	B	C	D		A	B	C	D
YES	3	1	4	4		3	2	4	1		3	1	3	3
NO	9	11	8	8		9	10	8	11		9	11	9	9

	CATEGORIES					CATEGORIES					CATEGORIES			
	A	B	C	D		A	B	C	D		A	B	C	D
YES	6	4	3	3		5	2	5	1		4	3	6	5
NO	6	8	9	9		7	10	7	11		8	9	6	7

	CATEGORIES					CATEGORIES					CATEGORIES			
	A	B	C	D		A	B	C	D		A	B	C	D
YES	3	2	4	3		1	5	4	4		5	3	5	3
NO	9	10	8	9		11	7	8	8		7	9	7	9

TABLE 4.10 MODER. MIX BINOMIAL COMPARISONS   SLIGHT DOSE RESPONSE

	CATEGORIES				CATEGORIES				CATEGORIES			
	A	B	C	D	A	B	C	D	A	B	C	D
YES	3	2	2	4	3	5	4	3	1	3	8	10
NO	9	10	10	8	9	7	8	9	11	9	4	2

	CATEGORIES				CATEGORIES				CATEGORIES			
	A	B	C	D	A	B	C	D	A	B	C	D
YES	3	7	2	5	5	5	10	5	4	5	5	9
NO	9	5	10	7	7	7	2	7	8	7	7	3

	CATEGORIES				CATEGORIES				CATEGORIES			
	A	B	C	D	A	B	C	D	A	B	C	D
YES	3	6	7	9	4	3	4	6	5	3	4	7
NO	9	6	5	3	8	9	8	6	7	9	8	5

	CATEGORIES				CATEGORIES				CATEGORIES			
	A	B	C	D	A	B	C	D	A	B	C	D
YES	3	2	5	7	3	3	4	7	1	2	5	8
NO	9	10	7	5	9	9	8	5	11	10	7	4

	CATEGORIES				CATEGORIES				CATEGORIES			
	A	B	C	D	A	B	C	D	A	B	C	D
YES	2	4	2	5	2	1	7	4	1	3	3	6
NO	10	8	10	7	10	11	5	8	11	9	9	6

	CATEGORIES				CATEGORIES				CATEGORIES			
	A	B	C	D	A	B	C	D	A	B	C	D
YES	2	6	5	5	3	3	7	2	4	6	2	3
NO	10	6	7	7	9	9	5	10	8	6	10	9

	CATEGORIES				CATEGORIES				CATEGORIES			
	A	B	C	D	A	B	C	D	A	B	C	D
YES	3	7	5	3	4	1	5	9	4	4	6	4
NO	9	5	7	9	8	11	7	3	8	8	6	8

	CATEGORIES				CATEGORIES				CATEGORIES			
	A	B	C	D	A	B	C	D	A	B	C	D
YES	5	6	4	6	6	7	5	8	3	1	7	6
NO	7	6	8	6	6	5	7	4	9	11	5	6

TABLE 4.11   MODER. MIX BINOMIAL COMPARISONS   MODERATE DOSE RESPONSE

	CATEGORIES A	B	C	D	CATEGORIES A	B	C	D	CATEGORIES A	B	C	D
YES	6	7	9	6	6	5	5	9	3	5	5	7
NO	6	5	3	6	6	7	7	3	9	7	7	5

	CATEGORIES A	B	C	D	CATEGORIES A	B	C	D	CATEGORIES A	B	C	D
YES	4	8	7	6	4	6	5	7	5	5	7	5
NO	8	4	5	6	8	6	7	5	7	7	5	7

	CATEGORIES A	B	C	D	CATEGORIES A	B	C	D	CATEGORIES A	B	C	D
YES	5	2	7	8	4	6	4	6	4	5	7	3
NO	7	10	5	4	8	6	8	6	8	7	5	9

	CATEGORIES A	B	C	D	CATEGORIES A	B	C	D	CATEGORIES A	B	C	D
YES	4	7	7	8	2	8	8	7	3	5	4	9
NO	8	5	5	4	10	4	4	5	9	7	8	3

	CATEGORIES A	B	C	D	CATEGORIES A	B	C	D	CATEGORIES A	B	C	D
YES	4	8	6	9	2	4	7	8	5	6	8	8
NO	8	4	6	3	10	8	5	4	7	6	4	4

	CATEGORIES A	B	C	D	CATEGORIES A	B	C	D	CATEGORIES A	B	C	D
YES	5	6	7	6	4	4	6	8	3	7	4	8
NO	7	6	5	6	8	8	6	4	9	5	8	4

	CATEGORIES A	B	C	D	CATEGORIES A	B	C	D	CATEGORIES A	B	C	D
YES	3	4	5	8	4	2	6	5	5	6	9	6
NO	9	8	7	4	8	10	6	7	7	6	3	6

	CATEGORIES A	B	C	D	CATEGORIES A	B	C	D	CATEGORIES A	B	C	D
YES	4	5	5	8	3	6	6	6	4	6	6	6
NO	8	7	7	4	9	6	6	6	8	6	6	6

TABLE 4.12 MODER. MIX BINOMIAL COMPARISONS   SEVERE DOSE RESPONSE

	CATEGORIES					CATEGORIES					CATEGORIES			
	A	B	C	D	A	B	C	D	A	B	C	D
YES	4	7	7	9	5	8	9	8	4	4	5	8
NO	8	5	5	3	7	4	3	4	8	8	7	4

	CATEGORIES				CATEGORIES				CATEGORIES			
	A	B	C	D	A	B	C	D	A	B	C	D
YES	1	6	4	11	2	5	8	9	3	7	7	9
NO	11	6	8	1	10	7	4	3	9	5	5	3

	CATEGORIES				CATEGORIES				CATEGORIES			
	A	B	C	D	A	B	C	D	A	B	C	D
YES	0	5	6	10	5	6	9	7	4	6	9	12
NO	12	7	6	2	7	6	3	5	8	6	3	0

	CATEGORIES				CATEGORIES				CATEGORIES			
	A	B	C	D	A	B	C	D	A	B	C	D
YES	6	7	9	9	2	6	6	10	1	8	4	8
NO	6	5	3	3	10	6	6	2	11	4	8	4

	CATEGORIES				CATEGORIES				CATEGORIES			
	A	B	C	D	A	B	C	D	A	B	C	D
YES	2	7	6	8	4	6	9	8	2	4	5	9
NO	10	5	6	4	8	6	3	4	10	8	7	3

	CATEGORIES				CATEGORIES				CATEGORIES			
	A	B	C	D	A	B	C	D	A	B	C	D
YES	4	8	7	11	2	5	9	7	0	4	6	10
NO	8	4	5	1	10	7	3	5	12	8	6	2

	CATEGORIES				CATEGORIES				CATEGORIES			
	A	B	C	D	A	B	C	D	A	B	C	D
YES	3	3	8	10	4	5	6	8	4	7	7	10
NO	9	9	4	2	8	7	6	4	8	5	5	2

	CATEGORIES				CATEGORIES				CATEGORIES			
	A	B	C	D	A	B	C	D	A	B	C	D
YES	5	5	11	9	7	7	10	11	4	3	7	9
NO	7	7	1	3	5	5	2	1	8	9	5	3

EXERCISES

Exercise 4.1: Estimation Dose Response--Browsing

Examine Table 4.1 and compare it to Table 4.4. Note the way in
which almost all contingency arrays in Table 4.4 show an increasing
frequency of "yes" counts from category A to category D. Can similar patterns be seen in Tables 4.1? Are there any "dose reversals"
(increases followed by decreases in counts of "yes" from left to
right) in Table 4.1? Compare Table 4.1 to Table 4.9. Can your eye
detect the reduction in variance? Compare Table 4.4 to 4.12. Can
the reduction in variance be seen in this comparison?

Exercise 4.2: Estimation Dose Response--Comparison of Estimates

For a binomial variate, the variance is a function of the underlying probability, and the variance of X/N as an estimator of p increases as p goes towards 0.5. This can be "adjusted for" by con-
sidering the random variable

    2 arcsin(sqr(X/N))

where the arcsin is in terms of radians. This function has an approximately constant variance of $1/N$.

The lack of constant variance for the initial estimator X/N
means that the precision of comparisons between two binomial vari-
ates decreases as the underlying p-values get closer to 0.5. Using
the arcsin transformation apparently eliminates this problem, but
the elimination is only illusionary. If $p1$ and $p2$ are two probabilities to be compared, a constant difference between them leads to a
large difference in the arcsin when they are both far from 0.5 and
to a smaller difference in the arcsin when they are both close to
0.5. If you think of a test of hypothesis about the difference between two p-values as a ratio of the difference being tested to its
standard error, then use of the untransformed variables produces a
ratio whose denominator increases as p approaches 0.5. Use of the

transformed variables produces a ratio whose denominator is constant
but whose numerator decreases as p approaches 0.5.

However, there are other advantages to using the arcsin transformation. If the probability of "yes" increases with dose, so that
the graph has a symmetric sigmoidal form (flat at the extremes but
sharply rising in the region around $p=0.5$), the arcsin transformation will appear to be a straight line through most of the p-values.
It is also easier to implement a blind computer program for the comparison of binomials by applying asymptotic normal theory to the
arcsin transformation, since this procedure does not involve the
problem of estimating the underlying p-values in order to compute
the asymptotic variance.

Another commonly used transformation of the binomial variate is
the logistic

$$\log (X/ (N-X))$$

If p is near 0 or 1 and if the underlying probability is a
function of some other variable (such as dose), the logistic transformation will often plot as a straight line. And, straight lines
(requiring only two parameters) are easier to handle than curved
ones.

Run this exercise for each of the tables in this chapter. Pick
three contingency arrays at random from a given table. Calculate
X/N, the arcsin and the logistic transformation for each category in
the first contingency array chosen. Then, plot these values on
graph paper against equally spaced "doses" A, B, C, and D. Add the
corresponding cells from the first two contingency tables (so each
N is now 24 instead of 12). Calculate the three estimates of effect
and plot them. Add the corresponding cells for all three contingency tables ($N=36$ for each category), and calculate and plot the
three estimators.

What effect does increasing sample size have on the regularity
of the resulting plots? Which plots seem more linear?

Exercise 4.3:   Two-by-Two Contingency Tables

If all else fails, most sets of data can be organized to yield two
proportions to be compared or some sort of contrasting two-way
classification of "yes" and "no" in each dimension. For this reason, the literature on 2 x 2 tables is immense. Much of it is
filled with arguments over whether tests should be "corrected" or
not, whether it is possible to construct confidence bounds on the
difference between probabilities associated with the two columns,
or whether it makes a difference if the total sum is fixed for the
columns (or the rows) or for both columns and rows together. Though
many of these discussions have merit in that they deal with real
problems of interpretation, we will not enter them here. Instead,
this exercise asks the student to look at 2 x 2 tables simply in
terms of the randomness of scatter.

This exercise should be run on each of Tables 4.1-4.12. Pick
five contingency arrays at random from a given table and copy down
the counts of "yes" and "no" for categories A and D only. To test
the hypothesis of independence of rows and columns (or, equivalently
equality of probability of a "yes" in both categories), you should
compute

1. A continuity corrected chi square test;
2. A chi square test without continuity correction;
3. A t-test of the equality of two proportions which treats
    each column as a set of N numbers, X of which are 1.0 and
    $N-X$ of which are 0.0;
4. A normal theory test of equality of two proportions using
    the arcsin transformation (see Exercise 4.2) and its
    asymptotic variance of $1/N$.

Compute these tests for the first of the tables chosen, for the
sum of the first two, for the sum of the first three, and for the
sum of all five. What effect can you see of

1.   Increasing sample size?
2.   Degree of "dose response"?
3.   Degree of mixing of binomials?

Exercise 4.4:   Testing Dose Response Versus More General Hypotheses

The power of a statistical test of hypothesis can be improved by
restricting the alternative hypotheses against which the null hypothesis is tested. An example of this occurs in the choice of
"one-sided" versus "two-sided" tests of the mean. If the mean being
tested can reasonably be expected to fall either greater or less
than the null hypothesis value, then the probability of a Type I
error has to be divided between two rejection regions. If there
is no good reason to believe that the random processes involved
would have a mean less than the null, then all the probability can
be concentrated in a rejection region with expectation greater than
the null mean, and a smaller degree of difference in the observed
data will provide significant evidence against the null hypothesis.
Similar opportunities arise in the examination of contingency tables.

 If we ignore the classifications into which a given contingency
table may be organized and consider it, instead, as a large collection of cells, then we can look at the contingency table in its most
general form. This is as a multinomial with M cells and M values of
p, which we might call $p1,p2,p3,...,pM$, describing the probability of
a given individual's falling into each cell. Since any individual
must fall into some cell the M values of $p1,p2,...,pM$ must all add up
to 1.0 and lie between 0.0 and 1.0. In this way, the most flexible
set of parameters that can describe the contingency table lie in an
M-1 dimensional space, called an M-simplex.

 If there are only two cells, then the set of p-values that will
meet these requirements can be graphed in two dimensions as in
Figure 4.1.

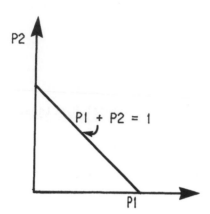

Fig. 4.1   The 2-Simplex

If there are three cells, then the set of p-values can be graphed in three dimensions as a flat triangular segment of a plane as in Figure 4.2.

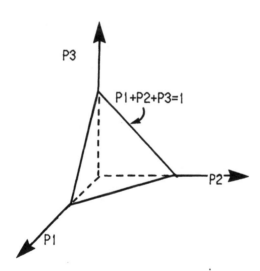

Fig. 4.2   The 3-Simplex

If there are more than three cells (as will occur in any contingency table of interest), the M-1 dimensional space of possible p-values (the M-simplex) is usually represented as a triangle. No one is expected to visualize such a simplex with the same dimensional clarity as that of the triangle that is the 3-simplex. However, there is a geometric terminology that is used to describe aspects of

the M-simplex, and the student should try to visualize the meaning
of this terminology. One way is to think of the 4-simplex (which is
a three-dimensional object) as a tetrahedron, looking a bit like one
of the pyramids of ancient Egypt. It has triangular faces (of two
dimensions) which are simplices themselves, edges (of one dimension)
which are also simplices, and vertices. One gets to a "face" of this
mixed simplex by requiring that one of the p-values be fixed at zero.
The face is then the simplex that results from allowing the remaining
p-values to range where they can. An "edge" is a face of one of the
faces of the larger simplex. If we deal with an M-simplex of great
dimension, we can think of a hierarchy of faces of faces of faces.
At any stage of dimension reduction, the currently considered simplex
has its faces, and the faces of these faces are called "edges."
There is only one set of vertices, however. These are the points
where all but one of the p-values are set to zero, and that one p-
value is fixed at 1.0.

There are, of course, other ways of reducing the dimensionality
of the M-simplex. Think again of the tetrahedron (the 4-simplex).
A slice through the tetrahedron will produce a triangle. That triangle need not be equilateral, like the faces of the tetrahedron,
but it will be a triangle, and it will be two-dimensional. Thus, it
is possible to construct spaces of lower dimension in the M-simplex
other than the faces.

An example can be seen in the standard statistical test for the
independence of rows and columns in a two-dimensional contingency
table. Suppose there are four columns and two rows (as in the
examples in this chapter). Without imposing any structure on the
table, it can be seen as a set of eight cells, and the set of underlying p-values lie in an 8-simplex (a space of 7 dimensions). If
we assume that the rows and columns are independent, then the probability of being in a given row and column cell is the product of the
probability of being in that row times the probability of being in
that column. There are thus six probabilities to be considered in
the null hypothesis, instead of eight. However, the row probabilities must all add to one, and the column probabilities must also add
to one. So, there is really one dimension in the rows and three

dimensions in the columns, and the hypothesis of row and column independence can be thought of as lying in a four-dimensional space,
imbedded in the 8-simplex.

(This four-dimensional space is not, itself, a simplex. There
is a three-dimensional simplicial "slice" of the outer 8-simplex
that represents the four probabilities for columns. "Over" this
subsimplex, we erect a set of one-dimensional simplexes that one can
visualize as an infinite number of thin sticks. If the reader's
geometrical imagination fails at this point, he or she should remember that this entire paragraph is a parenthetical aside. It is designed to keep this discussion correct in a mathematical sense, but
it is enclosed in parentheses to show that understanding it is not
necessary to an adequate understanding of the rest of the discussion.)

The usual test of row and column independence calculates a
"distance" from the observed point to the nearest point in the four-
dimensional space defined by the hypothesis of row and column independence. This is the chi square contingency table test described
in elementary textbooks. Actually, there are two "distance" measurements involved. That is, there are two chi square statistics calculated. The other statistic results from what the statistician
thinks is a reasonable set of alternative hypotheses.

In the entire M-simplex that describes all mathematically
possible p-values, there may be some that are impossible under the
practical conditions that lead to the contingency table. For instance, suppose we categorized a person's fingerprints in terms of
the number of open sworls and closed loops that are counted in the
thumbprint. There is room on one thumb for only a limited number of
ridges (look at your own thumb to confirm this). A thumb with a
large number of open sworls has little room for closed loops and
vice versa. Thus, a contingency table that has counts of sworls as
rows and counts of loops as columns must have a decreasing probability off the diagonal and zero probabilities in the extreme upper
right and lower left cells. Any reasonable test of hypotheses about
these counts would have to acknowledge that other patterns of probability should not be considered.

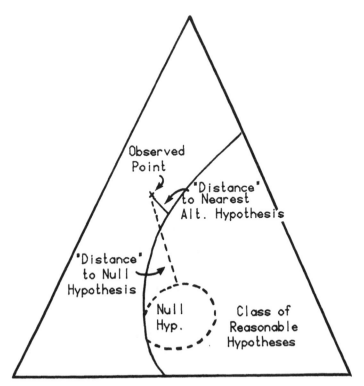

Fig. 4.3 Restricted Chi Square Test

Thus, in a practical situation, the p-value associated with some of the mathematically possible points in the M-simplex should be ignored. It may be conceivable that random noise will produce observed values of $X(i)/N$ that fall into such regions, but the analyst must consider them as impossible. This is analogous to a one-sided test of the mean, where it is possible for the observed average to be less than the hypothesized mean due to random noise but the only reasonable alternative to the null hypothesis is that the true mean is greater than the null hypothesis mean.

One can visualize this situation as in Figure 4.3.

Note that the null hypothesis to be tested is part of the class of reasonable alternative hypotheses. The region of the simplex assigned to the null hypothesis is a subset of the region assigned to the class of reasonable alternatives. One can now compute a chi

square "distance" from the observed point to the nearest point in
the class of reasonable hypotheses (the solid line in Figure 4.3)
and another chi square "distance" to the nearest point in the null
hypothesis (the dotted line in Figure 4.3). Since the null hypothesis is embedded in the class of reasonable hypotheses, the second
"distance" will always be greater than or equal to the first one.
The difference between these two statistics is called the Restricted
Chi Square Test and is due to work by Jerzy Neyman, but it is seldom
part of standard textbooks.

Neyman has shown that the asymptotic distribution of the Restricted Chi Square Test has a simple form under at least one
specific situation. If the class of reasonable hypotheses is part
of a space whose dimension is less than that of the M-simplex and if
the null hypothesis is part of a space whose dimension is less than
that of the class of reasonable hypotheses, then the Restricted Chi
Square Test has an asymptotic distribution of a chi square whose
degree of freedom is the difference between the dimensions of the
two subspaces.

Let us see how the standard test for independence of rows and
columns fits into this framework. The null hypothesis, as we have
seen, has a dimension equal to the sum of the number of rows and
columns, less two. In the 4 x 2 contingency table, the outer simplex has dimension 7 and the null hypothesis space has dimension 4.
If we let the entire outer 8-simplex (of dimension 7) be the class
of reasonable hypotheses, then the Restricted Chi Square Test statistic would be distributed as a chi square with 3 (=7-4) degrees of
freedom. If the set of alternative hypotheses is the entire 8-
simplex, then the nearest point in that class to the observed point
is the observed point, itself. The Restricted Chi Square Test is
then the usual chi square test for independence of rows and columns
less zero (the "distance" from the observed point to itself). Which
leads us to the proper result.

Suppose, now, that there is a restriction on the class of reasonable alternatives for the contingency tables in this chapter,
such that any change in probability of "yes" going from category A
to category D has to be nondecreasing. That is, suppose there is a

"dose response" imposed on all tables, such that the null hypothesis is that the "dose response" is flat and that all probabilities are equal from category A to category D. We imbed this null hypothesis in the class of reasonable alternatives that state that any difference in probability has to be increasing at some point or points from A to B to C to D. To construct the Restricted Chi Square Test, one needs to compute the point nearest the observed values in the class of all possible nondecreasing "dose responses." Next, we would compute the usual chi square test of independence of rows and column (which is the "distance" from the observed point to the nearest point in the null hypothesis). We subtract the first from the second and the result is our test statistic.

There is one problem with such a procedure. The set of reasonable hypotheses is not a subspace of the M-simplex of reduced dimension. So, Neyman's derivation cannot be used to find the distribution of this test statistic. Bartholomew has found the distribution of this test statistic. His distribution is calculated as a sum of chi squares plus a set of probabilities derived from Sterling numbers--all in all a very messy operation.

There is another way of getting at this problem. If we assume that the entire class of reasonable alternatives can be characterized by letting the p-values be a function of the doses with a small number of parameters, then the space of alternative hypotheses has dimension equal to that number of parameters. The null hypothesis can then be thought of as fixing some of those parameters at zero, and as having a dimension equal to the number of parameters left to be estimated from the data. For instance, we might assign the five categories arbitrary "doses" indicated by the variable D. We might then propose that the probability of a "yes" is a linear function of D, of the form:

Prob("Yes" for dose D) = A + BD

Then, under the null hypothesis, one of the parameters, B, is set at zero. The dimension of the class of reasonable alternatives is set at two. The dimension of the null hypothesis is one, and

the Restricted Chi Square Test has an asymptotic distribution of a
chi square with one degree of freedom.

Such a test has been in existence for many years. It was first
described in a published paper by Armitage, but it had been used by
others before that. It is usually called Armitage's test for linear
trend in proportions. However, it can be found in some textbooks
without Armitage's name attached. After both chi square "distances"
have been computed, it is possible, with a few lines of algebra, to
combine the two formulas, cancel out the portions that cancel, and
combine the portions that combine, yielding the following test
statistic:

Let

$n(i)$ = number of observations both "yes" and "no"

$N = \Sigma n(i)$

$X(i)$ = number of "yes" observations, ith category

$D(i)$ = the coded "dose" associated with ith category

$d = \Sigma n(i)D(i)/N$

$P(i) = X(i)/n(i)$

$p = \Sigma P(i)/k$, k=No. of doses

num $= \Sigma n(i)(P(i)-p)(D(i)-d)$

den $= p(1-p)/(\Sigma n(i)D(i)^2 - N\,d^2)$

$T = \text{num}^2/\text{den}$ is the required test statistic

With this tool, not usually found in elementary text books, the
student is ready for this exercise, which can be done on each of the
Tables 4.1-4.12.

Pick five contingency arrays at random from a given table. Run
the following tests, first on one array, then on the sum of the first
two arrays, then on the sum of the first three arrays, then on the
sum of all five arrays:

1. The usual chi square test of independence for rows and columns;
2. Armitage's test for coded "doses" A=1, B=2, C=3, D=4;
3. Armitage's test for coded "doses" A=1, B=2, C=4, D=8 (equal
    intervals on a log scale).

Examine the results of these tests to consider the effects of restriction of alternative hypotheses (general chi square test for independence versus Armitage's test), the arbitrary choice of coded doses, increasing dose response, and increasing degrees of mixing of binomials.

SELECTED WORKED-OUT EXAMPLES

Exercise 4.2, applied to Table 4.3:

Starting Point: column 2, row 4

	A	B	C	D
$n = 12$				
yes	4	3	4	3
no	8	9	8	9
$n = 24$				
yes	6	7	8	11
no	18	17	16	13
$n = 36$				
yes	10	10	10	11
no	26	26	26	25
Estimates of Prob(Yes):				
$n = 12$				
x/n	0.333	0.250	0.333	0.250
Arcsine tr.	1.231	1.047	1.231	1.047
Logit	-0.693	-1.099	-0.693	-1.099
$n = 24$				
x/n	0.250	0.292	0.333	0.458
Arcsine tr.	1.047	1.141	1.231	1.487
Logit	-1.098	-0.887	-0.693	-0.167
$n = 36$				
x/n	0.278	0.278	0.278	0.306
Arcsine tr.	1.110	1.110	1.110	1.171
Logit	-0.956	-0.956	-0.956	-0.821

Exercise 4.3 applied to Tables 4.9-4.12:

Starting Points

Table	col	row
4.9	2	4
4.10	3	5
4.11	1	4
4.12	2	7

Table	n	$X/(n-X)$ A	$X/(n-X)$ D	Chi Squares cor.	Chi Squares uncor.	t-test	Normal z-test
4.9	12	0/12	1/11	0.00	1.04	1.48	1.48
	36	5/31	6/30	0.00	0.11	0.46	0.32
	60	8/52	10/50	0.07	0.26	0.72	0.50
4.10	12	1/11	6/6	3.23	5.04	3.57	2.53
	36	4/32	12/24	3.94	5.14	3.33	2.33
	60	11/49	20/40	2.78	3.52	2.69	1.82
4.11	12	1/11	4/8	1.01	2.278	2.24	1.62
	36	6/30	12/24	1.85	2.67	2.35	1.61
	50	9/51	18/24	3.06	3.87	2.83	1.94
4.12	12	2/10	4/8	0.22	0.89	1.36	0.94
	36	8/28	18/18	4.88	6.02	3.63	2.42
	60	13/47	31/29	10.37	11.63	5.07	3.36

Note

1. All four tests are in general agreement with respect to significance.

2. The degree of agreement increases as n increases.

3. There is an increase in power going from Table 4.9 to Tables 4.10, to 4.11, to 4.12. However, the first random choice in Table 4.12 ($n=12$) fails to come up significant.

Exercise 4.4, applied to Table 4.4:

Starting Point: column 1, row 6

	A	B	C	D
$n = 12$				
yes	3	3	6	7
no	9	9	6	5
$n = 36$				
yes	5	11	13	21
no	31	25	23	15
$n = 60$				
yes	12	19	20	36
no	48	41	40	24

		Armitage Tests					
	Chi	Linear Coding			Log Coding		
N	Square	num	den	test	num	den	test
12	4.44	0.62	0.10	3.92	1.48	0.57	3.82
36	16.05	0.69	0.03	15.32	1.65	0.46	5.96
60	22.27	0.61	0.02	19.32	1.53	0.11	21.12

significance level cut-off values:

5%	7.81			3.85			3.85
1%	11.34			6.63			6.63

Note how the greater power of the restricted tests shows up for the smallest sample size, and note how use of linear versus log coding makes little difference with respect to findings of significance.

5

MODELS VERSUS DATA

INTRODUCTION

A statistician is frequently called upon to estimate the parameters
of a simple model from a set of data. The inexperienced statistician can be fooled by the random fall of such data, detecting nonlinear relationships where none exist, finding "peculiar" outliers
which are only part of the random noise that might be expected. The
exercises in this chapter are designed to provide a "feel" for what
might be expected when well-behaved random errors are added to
simple models.

There are two sets of tables in this chapter. Tables 5.1, 5.2,
and 5.3 compare two regression models:

$$Y(1) = 10 + x + E, \text{ and}$$
$$Y(2) = 8 + \min(x,5) + E$$

where

$x = 1, 2, 3, 4, 5, 6,$ and
E is normally distributed with mean 0

TABLE 5.1 LINEAR(Y1) VRS NON-LINEAR(Y2)--SLIGHT VARIANCE

X	Y1	Y1	Y1	Y1	Y1	Y2	Y2	Y2	Y2	Y2
1.	9.9	10.1	9.4	9.2	9.2	9.1	7.3	8.4	8.9	8.2
2.	9.3	11.2	12.8	13.2	12.0	8.2	9.4	10.1	12.3	11.7
3.	11.1	9.9	13.0	12.4	13.2	12.4	11.3	12.2	10.9	12.8
4.	12.9	11.0	11.3	12.9	15.7	12.7	12.6	11.4	11.5	13.0
5.	15.5	15.8	16.3	13.5	17.2	9.0	9.1	7.6	5.9	9.3
6.	15.7	18.0	17.9	13.9	17.0	5.8	5.8	6.4	7.6	5.8
1.	10.7	9.4	11.5	10.7	12.8	6.6	11.4	9.1	8.7	8.7
2.	13.8	13.3	12.8	13.3	9.9	10.4	11.9	10.2	11.4	9.4
3.	14.7	12.8	16.1	12.9	15.5	8.9	10.7	11.8	9.2	10.3
4.	17.1	13.7	17.0	17.1	14.2	11.7	13.3	14.9	11.0	11.8
5.	15.7	15.5	15.1	14.8	13.5	7.6	10.5	10.9	10.3	10.2
6.	16.0	16.6	18.9	16.1	16.4	10.6	7.6	10.1	9.0	8.5
1.	11.0	10.7	9.8	10.2	10.0	9.3	11.5	5.9	10.7	11.6
2.	9.7	11.8	10.4	10.7	12.0	9.9	10.0	8.5	11.4	9.3
3.	13.6	11.9	15.0	11.7	15.9	14.0	11.4	10.2	10.6	10.0
4.	13.4	14.1	15.1	14.4	13.6	11.8	10.3	12.3	11.2	9.5
5.	14.2	16.6	17.0	16.7	16.8	7.4	7.0	10.4	6.5	10.8
6.	17.8	18.8	12.9	13.0	13.0	5.2	4.9	8.2	8.2	6.1
1.	11.5	8.0	9.1	10.8	10.9	9.0	6.2	9.5	9.0	11.2
2.	11.9	13.9	12.1	10.3	15.1	8.4	11.1	11.2	9.7	8.9
3.	13.6	11.0	12.4	11.7	13.1	11.2	12.1	10.5	12.3	9.7
4.	13.8	13.0	12.9	16.7	12.4	11.1	9.9	15.0	12.8	13.1
5.	16.4	15.3	13.6	14.8	14.1	10.6	10.4	8.9	8.5	6.1
6.	13.4	15.3	16.1	15.6	15.1	5.0	7.3	10.7	9.1	8.4

TABLE 5.2 LINEAR(Y1) VRS NON-LINEAR(Y2)--MODERATE VARIANCE

X	Y1	Y1	Y1	Y1	Y1	Y2	Y2	Y2	Y2	Y2
1.	13.6	10.9	5.4	10.6	11.4	4.6	4.0	9.4	3.6	7.9
2.	19.2	12.8	7.0	19.0	14.7	15.5	8.5	5.3	12.2	5.7
3.	10.3	12.0	13.0	11.3	12.8	10.5	10.7	5.9	14.4	9.3
4.	19.0	18.2	13.2	12.3	8.7	17.5	8.9	17.4	12.1	12.2
5.	14.6	11.2	13.7	20.2	14.9	2.7	3.0	6.5	7.5	6.8
6.	15.1	13.2	12.3	18.0	16.9	14.7	6.1	13.0	12.4	4.7
1.	5.3	16.1	14.3	10.7	12.3	9.9	11.5	11.7	12.6	10.5
2.	14.3	15.0	17.1	13.1	11.8	7.4	4.9	5.0	2.5	9.1
3.	15.1	8.4	9.5	12.4	15.8	6.8	4.9	5.2	17.1	14.7
4.	14.1	17.7	16.4	16.8	8.5	9.5	7.1	13.1	9.7	7.9
5.	14.5	15.5	17.0	9.5	13.3	6.4	2.5	2.6	1.8	12.7
6.	16.1	15.2	15.0	19.1	13.9	7.4	6.5	10.5	6.1	8.7
1.	12.4	5.3	11.2	15.0	5.2	6.2	11.6	9.8	5.3	10.4
2.	12.6	17.6	18.1	17.5	5.2	14.2	10.2	6.8	8.8	17.0
3.	15.0	11.0	12.7	10.9	15.0	12.3	12.4	11.1	6.9	14.4
4.	14.0	12.8	10.5	18.8	14.1	8.5	10.4	10.6	14.4	10.6
5.	17.2	11.7	21.8	13.5	13.2	12.5	6.8	14.4	5.3	6.5
6.	11.4	16.2	18.1	13.0	13.3	1.3	4.7	3.6	5.1	10.6
1.	7.6	6.3	13.8	11.0	7.3	11.3	9.4	14.7	7.0	13.7
2.	13.0	13.0	13.6	6.0	9.6	6.7	15.5	10.5	8.4	11.8
3.	14.8	12.2	13.7	9.2	18.9	10.3	12.3	11.3	11.4	9.2
4.	21.5	19.7	13.0	16.7	11.6	11.9	18.7	4.4	8.5	10.4
5.	16.0	7.7	11.2	13.3	16.2	12.1	4.3	2.9	4.1	8.0
6.	16.6	17.4	11.5	16.0	20.4	5.9	6.6	5.0	3.5	8.1

TABLE 5.3 LINEAR(Y1) VRS NON-LINEAR(Y2)--GREAT VARIANCE

X	Y1	Y1	Y1	Y1	Y1	Y2	Y2	Y2	Y2	Y2
1.	3.6	1.8	9.4	16.6	4.8	5.1	13.5	5.6	8.3	17.3
2.	-0.5	15.7	13.0	14.6	9.0	24.1	-2.5	16.9	-4.5	3.8
3.	17.9	23.8	9.6	10.4	0.3	6.4	2.9	18.0	15.9	2.9
4.	6.4	15.0	27.0	12.1	11.9	10.0	13.9	24.9	10.1	12.1
5.	6.1	16.9	9.5	27.1	3.3	18.7	16.4	9.6	-1.7	7.7
6.	26.2	1.5	12.5	13.1	12.0	22.3	11.3	6.4	10.3	2.1
1.	10.1	8.5	5.6	24.3	19.4	24.0	8.9	12.7	5.5	-6.4
2.	8.5	-2.8	11.3	7.2	2.3	10.9	11.2	11.5	11.7	5.3
3.	14.3	11.9	28.2	0.8	27.1	7.6	-1.3	2.4	8.2	0.1
4.	20.9	13.0	9.9	19.8	15.8	14.0	21.2	-2.5	25.6	14.9
5.	10.0	14.4	2.6	15.6	23.1	4.1	21.5	12.2	9.2	10.1
6.	16.7	16.5	15.1	21.6	21.8	-6.8	-6.3	11.9	21.6	18.6
1.	2.8	17.9	4.7	13.5	19.6	17.0	19.3	11.4	5.2	13.0
2.	15.1	26.0	10.3	13.6	22.2	19.4	8.8	1.9	13.1	9.6
3.	15.6	1.1	0.6	19.2	27.9	6.8	11.4	10.1	11.4	17.3
4.	16.6	16.4	13.9	4.4	27.7	1.5	6.2	13.1	14.0	14.5
5.	11.4	29.0	12.5	13.9	16.4	19.4	4.1	-7.5	14.3	6.4
6.	8.0	12.3	15.6	11.8	15.4	8.3	18.5	11.0	-0.3	4.7
1.	0.3	20.2	8.5	13.9	7.0	8.9	6.3	14.2	-1.2	12.5
2.	8.5	19.8	13.1	18.1	25.6	15.5	20.6	10.0	-2.1	-5.5
3.	3.7	-0.2	15.3	20.7	-1.2	0.1	11.4	11.0	24.4	19.3
4.	17.2	28.8	11.2	16.7	-0.6	7.8	15.9	26.9	8.8	3.0
5.	28.7	17.4	13.6	14.3	12.3	7.6	9.6	17.2	7.1	12.7
6.	0.4	5.8	24.7	22.6	15.9	7.7	-1.4	11.4	10.9	17.6

Tables 5.4, 5.5, and 5.6 display realizations of the sigmoidal
regression:

$Y = 8 + E$ if x is less than 3

$Y = 5 + x + E$ if x is between 3 and 7

$Y = 12 + E$ if x is greater than 7

where

$x = 1, 2, 3, 4, 5, 6, 7, 8, 9$, and

E is normally distributed with mean 0

TABLE 5.4 SIGMOID REGRESSION--SLIGHT VARIANCE

X	Y	Y	Y	Y	Y	Y	Y	Y
1.	5.5	7.8	8.9	8.2	5.3	8.5	7.1	8.9
2.	7.8	5.5	7.2	8.7	8.6	7.8	10.0	5.5
3.	8.5	8.1	6.7	10.2	7.6	7.7	6.9	8.4
4.	9.0	9.4	11.5	7.4	8.7	8.1	9.6	9.4
5.	8.8	9.8	7.8	10.8	8.9	10.1	9.1	10.2
6.	13.1	13.2	12.0	10.2	11.4	10.5	13.6	10.9
7.	10.7	11.6	11.1	12.0	11.5	13.5	11.8	11.1
8.	9.9	11.2	9.3	10.6	9.5	12.7	11.9	11.7
9.	10.9	12.9	13.5	13.9	10.4	11.3	13.2	10.6
1.	9.7	8.7	6.0	9.9	8.3	7.6	10.8	7.5
2.	7.8	9.4	9.0	9.5	8.0	9.0	9.9	10.2
3.	6.5	5.5	8.1	8.0	9.6	10.1	10.6	7.9
4.	7.9	8.1	11.8	7.5	8.5	8.8	11.6	8.9
5.	12.3	11.0	10.8	10.0	10.3	8.7	10.8	9.8
6.	10.7	11.9	11.0	10.7	12.0	10.1	10.5	11.4
7.	12.0	11.9	11.7	12.1	11.7	12.4	13.8	11.0
8.	9.4	12.2	10.8	11.7	12.6	13.8	12.0	11.8
9.	11.6	11.6	12.5	9.2	11.9	10.2	13.7	14.2
1.	8.3	7.9	9.1	7.1	8.6	5.2	7.4	9.2
2.	7.1	6.1	8.4	8.3	6.0	6.0	6.3	8.2
3.	5.4	8.7	6.7	9.0	9.4	8.7	7.7	9.3
4.	6.7	11.4	10.2	9.9	8.9	7.7	7.6	6.7
5.	12.8	9.2	12.6	9.9	9.8	10.0	10.7	8.5
6.	11.7	11.6	9.4	9.9	9.0	10.7	10.9	9.7
7.	11.4	12.1	14.6	11.3	10.3	11.5	10.9	14.2
8.	11.9	11.8	11.9	13.1	11.2	11.4	13.4	11.6
9.	13.0	12.6	9.8	13.2	12.7	14.7	11.8	10.6

TABLE 5.5 SIGMOID REGRESSION--MODERATE VARIANCE

X	Y	Y	Y	Y	Y	Y	Y	Y
1.	10.9	9.1	12.0	10.2	6.4	5.3	8.0	7.0
2.	12.8	9.3	7.0	12.9	2.2	10.1	4.8	9.6
3.	5.1	8.8	13.4	13.9	9.5	11.2	8.3	11.8
4.	10.0	14.2	7.4	10.3	10.2	7.2	8.9	3.8
5.	7.4	8.1	14.2	10.5	12.7	10.0	12.9	11.3
6.	10.8	15.2	7.4	8.6	14.2	11.0	8.7	14.6
7.	18.2	10.8	9.5	6.3	13.1	10.7	10.8	12.6
8.	7.8	11.5	15.3	13.0	12.3	8.8	11.5	12.9
9.	7.0	8.2	12.4	12.0	6.3	16.2	12.2	9.2
1.	8.8	13.5	10.0	13.4	5.5	3.4	12.2	5.4
2.	3.3	3.2	8.0	7.1	7.9	7.8	5.9	10.2
3.	14.0	13.1	10.1	5.8	8.4	11.3	13.5	8.3
4.	10.7	8.3	10.1	9.6	14.3	13.2	6.2	5.5
5.	8.3	7.8	14.2	11.5	9.9	10.7	13.7	15.7
6.	10.4	6.9	8.2	6.2	4.9	11.2	5.2	14.5
7.	11.0	11.0	13.4	8.1	12.5	16.9	10.5	11.5
8.	16.5	11.5	6.3	12.6	9.3	13.2	16.5	12.5
9.	11.8	11.1	8.3	10.3	10.9	16.3	11.8	12.5
1.	6.5	4.4	9.8	7.6	10.6	11.4	6.5	2.9
2.	6.4	11.8	12.3	10.7	10.1	5.8	4.1	6.8
3.	9.3	10.8	8.8	6.8	5.3	13.2	3.7	7.4
4.	12.0	5.1	9.7	12.7	3.3	8.7	13.9	6.1
5.	9.0	9.5	10.0	11.2	5.5	11.6	15.9	9.5
6.	7.9	12.4	10.1	11.8	8.5	9.8	12.4	9.2
7.	6.2	9.4	11.0	5.8	8.0	10.8	6.7	8.6
8.	18.1	16.8	17.5	12.2	14.0	8.2	13.3	12.4
9.	10.1	11.9	16.2	10.8	16.1	11.1	7.6	14.8

TABLE 5.6 SIGMOID REGRESSION--GREAT VARIANCE

X	Y	Y	Y	Y	Y	Y	Y	Y
1.	-5.3	5.1	12.7	-3.8	7.8	-5.1	10.8	16.2
2.	4.9	5.0	5.6	6.4	-5.6	10.1	5.6	5.6
3.	6.6	15.3	2.8	18.1	-5.6	9.8	-0.4	2.6
4.	3.8	5.3	7.5	11.7	19.4	8.4	13.8	5.7
5.	-0.5	4.0	15.4	10.6	4.6	7.3	11.8	5.7
6.	8.1	16.2	19.0	19.4	12.3	15.8	7.9	13.1
7.	6.0	3.8	15.1	14.1	17.3	18.5	12.0	9.9
8.	5.2	21.8	19.3	25.0	13.1	14.1	18.8	7.5
9.	12.9	5.4	9.7	16.3	11.4	11.6	12.4	21.1
1.	21.4	4.2	13.5	7.2	5.2	10.1	14.4	6.2
2.	5.4	-0.9	-0.5	12.8	13.2	11.8	8.6	5.2
3.	9.3	15.2	8.9	10.8	10.8	7.6	7.6	-0.3
4.	12.0	-1.8	20.6	22.6	4.4	8.3	-0.4	15.1
5.	19.4	6.0	12.6	9.8	-1.4	10.0	10.4	-3.7
6.	23.2	24.5	16.1	21.2	11.8	6.0	18.6	19.7
7.	3.5	14.3	20.5	10.0	12.1	12.8	9.1	19.4
8.	18.2	5.6	2.9	8.7	19.6	12.0	18.7	16.3
9.	11.6	8.0	9.5	13.4	0.3	11.7	9.4	24.8
1.	18.2	9.9	11.1	10.1	10.2	11.3	3.1	16.6
2.	13.5	3.5	6.0	14.0	6.1	21.9	17.4	15.7
3.	15.1	17.5	14.7	11.2	3.5	3.2	20.8	9.0
4.	13.9	10.2	2.4	2.2	9.5	6.6	4.3	7.3
5.	9.8	8.0	21.5	3.1	11.8	6.8	13.7	18.6
6.	-0.6	13.8	11.7	14.3	12.0	0.4	9.4	15.4
7.	18.2	8.6	7.9	25.0	8.0	14.0	8.6	17.8
8.	5.1	13.4	18.8	22.4	14.1	14.0	11.9	6.2
9.	8.9	20.7	0.3	14.2	2.4	-0.7	13.2	0.4

The variance of E increases going from Table 5.1 to 5.3 and going from Table 5.4 to 5.6.

The first set of tables will allow the student to compare a steadily increasing linear response to one that flattens out for large values of the independent variable. The second set of tables will allow the student to search for the linear portion of a response curve which is flat on both ends.

EXERCISES

Exercise 5.1:   Browsing Among Tables 5.1-5.3

Glance through Tables 5.1-5.3, comparing the patterns of values for $Y(1)$ (linear) to those of $Y(2)$ (nonlinear). Compare Table 5.1 (slight variance) to Table 5.3 (great variance), to see how badly an additive random error can confuse the issue. How often do the patterns of Y-values appear to fit the underlying regressions involved? Is it possible to identify extreme values and eliminate or reduce their influence?

Exercise 5.2:   Comparing Linear to Nonlinear

Run this exercise once for each of Tables 5.1, 5.2, and 5.3. Each table is organized around sets of x-values 1, 2, 3, 4, 5, and 6. Pick one such set at random and a random starting column for $Y1$. Pick another random starting column for $Y2$. Copy out five sets of $Y1$ and $Y2$ data, continuing on to the next set of x-values or wrapping around to the beginning of the table, if necessary.

Plot single sets of $Y1$ and $Y2$ against the appropriate x-values. Calculate the least squares linear fits for each set and plot the appropriate lines. For the $Y1$ sets, the slope of the fitted line should be around 1.0 and the residual scatter should fall in random patterns about that line. For the $Y2$ sets, the slope will tend to be less than 1.0, and the points will tend to be above the line in the middle range of x-values. How well can such patterns be seen in single sets of Y-values?

Plot sets of three Y1 values for each x and three Y2 values for each x Look to see if there is more regularity to the residual patterns about the least squares lines. Mark the median Y-values for each x-value. How well do these medians fit the least squares lines? How well do they display the different underlying models?

Plot all five sets of Y1 and Y2 against the appropriate x-values, the least squares straight lines, and the median Y values for each x-value. Examine the regularity under these circumstances. What has been gained by using five rather than three Y-values for each x?

Exercise 5.3: Browsing Among Tables 5.4-5.6

Tables 5.4, 5.5, and 5.6 display random patterns about a simple sigmoidal regression. The regression is constant for x = 1, 2, or 3, rises linearly for x = 4, 5, or 6, and remains constant for x = 7, 8, or 9. Compare Table 5.4 (slight variance) to Table 5.5 (moderate variance) to Table 5.6 (great variance). How well does this pattern emerge through the increasing clouds of variability? See how many of the sets of Y-values appear to follow the sigmoidal pattern? How many appear to be linear throughout the range of x-values? For how many is it possible to estimate the "break-points" on either side of the linear region?

EXERCISE 5.4: Hunting for the Linear Portion of a Curve

This exercise should be run once for each of Tables 5.4, 5.5, and 5.6. Each table is organized around three sets of x-values from 1 to 9. Pick one of the sets of x-values at random and a column of Y-values also chosen at random. Copy out five sets of Y-values, continuing on to the next set of x-values or wrapping around to the beginning of the table, if necessary. Plot first one set of Y-values, then three sets, then all five sets against the appropriate x-values.

After plotting, compute the least squares linear regressions for the entire set of points. For low values of x, the points will

tend to lie above the line; for high values of x, the points will
tend to lie below the line. Note how the least squares line is influenced by random errors at the extremes of the x-range. When
plotting the three sets of Y-values, mark the medians and note how
they do or do not follow the sigmoidal patterns that should be
there. Do the same when plotting the five sets of Y-values.

Can you think of a way you might try to estimate the break-
points of the sigmoidal curve or test the hypothesis that the break-
points occur at $x=3$ and $x=7$ (this is not a trivial problem in
mathematical statistics)? Can you think of some useful decision
rule for eliminating points that do not fall in the linear portion
of the curve? One such rule might be to eliminate points at the
extreme range of the x-values, as long as the Y-values are in the
lower or upper quartiles (as appropriate for the end of the curve
being examined). How well does such a rule perform? Can you improve
upon it?

SELECTED WORKED-OUT EXAMPLES

Exercise 5.2, applied to Table 5.2:

Starting point row (group) = 2
                    $Y1$ Column=5, $Y2$ Column=3

x	$Y1$	$Y1$	$Y1$	$Y1$	$Y1$
1	12.3	12.4	5.3	11.2	15.0
2	11.8	12.6	17.6	18.1	17.5
3	15.8	15.0	11.0	12.7	10.9
4	8.5	14.0	12.8	10.5	18.8
5	13.3	17.2	11.7	21.8	13.5
6	13.9	11.4	16.2	18.1	13.0

x	Y2	Y2	Y2	Y2	Y2
1	11.7	12.6	10.5	6.2	11.6
2	5.0	2.5	9.1	14.2	10.2
3	5.2	17.1	14.7	12.3	12.4
4	13.1	9.7	7.9	8.5	10.4
5	2.6	1.8	12.7	12.5	6.8
6	10.5	6.1	8.7	1.3	4.7

Least Squares Lines of the form $Y(i)=A+Bx(i)$ + error

	n	using all data		using medians	
		A	B	A	B
Y1	1	12.08	0.15	-----	----
	3	11.21	0.49	12.53	0.42
	5	13.16	0.19	14.17	-0.11
Y2	1	8.55	-0.15	-----	----
	3	10.72	-0.50	11.45	-0.78
	5	11.73	-0.76	12.99	-1.06

Note how the estimates of A and B are not very close to the true values. However, the relationship that B for Y1 is greater than B for Y2 does emerge.

Exercise 5.4, applied to Table 5.4:

Start in row (group) 1, column 4

x			Y-values		
1	8.2	5.3	8.5	7.1	8.9
2	8.7	8.6	7.8	10.0	5.5
3	10.2	7.6	7.7	6.9	8.4
4	7.4	8.7	8.1	9.6	9.4
5	10.8	8.9	10.1	9.1	10.2
6	10.2	11.4	10.5	13.6	10.9
7	12.0	11.5	13.5	11.8	11.1
8	10.6	9.5	12.7	11.9	11.7
9	13.9	10.4	11.3	13.2	10.6

Least Squares Lines of the form $Y(i)=A+Bx(i)$ + error

n	using all data		using medians	
	A	B	A	B
1	7.31	0.58	----	----
2	6.77	0.60	7.23	0.49
3	6.88	0.59	7.21	0.54

Note how difficult it is to detect the flattening out of the sigmoid at either end from a single set of data. There are random
dose reversals, Y-values associated with the linear portion of the
curve look as "flat" as the Y-values associated with the ends.
However, a clear sigmoidal effect can be seen in the medians of
three and five values.

6

REAL LIFE DATA

INTRODUCTION

In the previous chapters, the student was given a sort of statistical pablum. The data were generated on a computer to provide examples of homogeneous random structures, each with a single well-defined aberration. "Real life" does not present the statistical analyst with such nice data. In fact, the working statistician is most often confronted with problems where the hypotheses to be tested and the parameters to be estimated are not even well defined, where he or she must decide whether the apparent aberrations in the data are essential to the solution of the problem or whether they are irrelevant nuisances that must be adjusted for.

The student in a formal statistics course is often warned to examine residuals and to watch out for outliers. However, it has been my observation that the novice is often led astray by purely random events that "look peculiar" once he or she leaves the solid rock of averages and other central tendencies. It is hoped that the previous chapters of this book have taught the student how peculiar purely random events do, in fact, look. Now, the guideposts are down, the real dice (with their unknown imperfections in weight) have been cast. Welcome to the uncharted world of random structures in real life!

EXERCISES

Exercise 6.1: Digit Preference

Nurses and physicians are taught to read blood pressure measurements
on a sphygmomanometer in terms of the nearest calibration mark and
never to interpolate values. The following data were derived from
blood pressures taken by a single individual using a device gradu-
ated in even mm of mercury. Thus, all recorded blood pressures
ended in 0, 2, 4, 6, or 8. The blood pressures were arranged in
groups of five, usually taken on the same day from five different
patients. There were 55 such groups (or 275 measurements, each, of
diastolic and systolic blood pressures). If the distribution of the
final digit were purely random, there is a probability of 1/5 that a
given reading will end in zero, and the number of zero endings in
five observations should be a binomial variate with $n=5$ and $p=0.20$.
Test this hypothesis for the systolic and diastolic measurements,
separately.

Since the measurements were all taken by the same individual,
there is a good chance that he has an underlying preference for or
against the use of zeros. This is the phenomenon of digit preference. If so, there should be some evidence that the probability of
having a reading ending in zero is different from 1/5. Test this
hypothesis. Can such a test be run without assuming that the data
are binomially distributed? If not, what evidence is there for or
against that more general assumption?

Frequencies of zero as last digit in sets of five blood pressures recorded by the same individual:

Number of times 0 occurred in a set of five	Observed Frequency	
	Diastolic	Systolic
0	30	21
1	13	18
2	9	8
3	3	7
4	0	1
5	0	0

Exercise 6.2:   IQ Versus Birth Conditions

One hundred five premature children born during a 2-year period were
classified by weight as a function of term into those with "small
birth size" and those with "appropriate birth size."   Eight years
later, their IQs were measured. The following table lists the IQs,
classified into intervals of 5 units each.   Is there any evidence
that, for such children, IQ at 8 years is related to size at birth?
For a "normal" population, IQ scores have a Gaussian distribution
with mean 100 and standard deviation 15.   Do either or both of these
groups of children have IQs that do not fit such a distribution?

Distribution of IQ scores as a function of birth size:

IQ	Observed Frequencies for Size	
	"Small"	"Appropriate"
less than 55	1	0
55-69	4	2
70-84	7	13
85-99	6	21
100-114	12	23
115-124	3	10
more than 124	0	3

Exercise 6.3:   Frequencies of Occurrence of Contentless Words

The Life of Davy Crockett, published by the Keystone Publishing Co.
(1889) is a combined printing of three works supposedly written by
Crockett, himself:

> *A Narrative of the Life of David Crockett; of the State
> of Tenn.* (1839);
>
> *An Account of Colonel Crockett's Tour to the North and
> Down East* (1835);
>
> *Colonel Crockett's Exploits and Adventures in Texas*
> (1836).

The first two were probably written by Crockett (with some
editing, it is thought, on the *Narrative*). The *Texas* book (which

includes the most detailed account available of the siege of the
Alamo) is believed to be a forgery.

The following are counts of appearance of the words "of" and
"as" per 500 words from the first 7500 words of each work. Test the
hypothesis that the mean frequencies are the same for a given word
for each work. Assuming that the *Narrative* and the *Tour* are written
by the same man, estimate whether the differences between both of
these and the *Texas* book are more than would be expected between the
Narrative and the *Tour*.

Note: The frequencies of contentless words can often be fit to
Poisson distributions.

Frequencies per 500 words:

Word Set	Narrative of	as	Tour of	as	Texas of	as
1	2	1	6	7	13	2
2	4	4	7	3	15	1
3	0	5	12	4	10	2
4	2	3	6	3	17	4
5	2	6	4	7	15	2
6	1	1	9	4	16	1
7	0	0	4	5	7	0
8	1	4	2	2	6	4
9	5	5	4	1	7	3
10	2	2	7	4	5	6
11	4	3	10	2	11	0
12	7	3	9	2	8	2
13	5	5	7	6	8	2
14	5	5	7	1	10	0
15	3	3	3	3	9	1

Exercise 6.4: Median Counts of Variable Macrophage Cells

Mouse macrophage cells were transferred to four tubes for each of
the treatments and doses indicated. After a predetermined period
of time, labeled "early," two tubes each were washed of treatment
and stained. Five different microscopic fields were chosen at
random for each tube and the median count of viable cells from the
five fields recorded. After a period of time, labeled "late," the

remaining two tubes at each treatment and concentration were washed
and stained, and the median count of viable cells from five fields
recorded.

The amount of time labeled "early" or "late" differed among the
treatments but was constant for all concentrations (including the
zero or controls) for any treatment.

For which of these treatments is there evidence of dose related cytotoxicity? Does the time of exposure (early versus late)
make a difference? Can you estimate the dose response for those
treatments that have had an effect?

Median counts of viable macrophage cells:

Treatment	Time	Concentration (mg/L)				
		0	1	0.1	0.01	0.001
Heparin	early	129	a	106	348	324
		476	543	531	437	479
	late	46	336	230	68	80
		412	104	94	376	194
Dextran	early	486	370	312	b	700
		401	442	512	424	606
	late	364	383	402	634	607
		536	357	428	594	588
Carageenum	early	666	170	351	426	656
		386	457	500	296	304
	late	733	200	337	294	234
		601	196	277	395	345
Sodium Cellulose	early	302	256	262	137	303
		283	129	154	345	145
	late	326	21	3	145	135
		296	8	6	11	54

[a]Test tube lost.
[b]All cells died within 24 hours of treatment.

Exercise 6.5: Asthmatic Attacks by Day of Week

Fifteen male patients with full-time jobs, who suffered from extrin-
sic asthma, were followed for 12 complete weeks, each, during their

asthma season. The number of asthmatic attacks was recorded by day
of week (there could be more than one attack on a given day), and
the total number of attacks by day for the 12-week period is tabu-
lated below.

There is a widely held medical conjecture that the number of
attacks is a function of the work week. One group holds that the
stress of work causes a build-up of attacks from Monday through
Friday, which then reduces over the weekend. Another group believes that the greatest number of attacks occur on Monday, with a
slightly fewer number on Tuesday, indicating the shock of return to
work, which then wears off. Still a third group believes that the
attacks are worse on weekends. Can any of these three hypotheses
be supported by this data?

Number of asthmatic attacks by day of week--12 weeks:

Patient	SUN	MON	TUE	WED	THU	FRI	SAT
1	4	5	5	4	4	1	1
2	8	5	10	8	7	6	7
3	0	1	2	1	1	2	1
4	7	7	10	2	4	1	5
5	3	8	6	4	7	5	9
6	3	0	0	6	12	9	0
7	2	5	5	10	1	1	4
8	2	3	8	4	9	9	5
9	0	2	7	9	9	6	10
10	1	2	0	1	9	7	2
11	4	9	0	10	3	8	2
12	5	0	8	9	6	2	4
13	3	5	3	4	7	6	4
14	5	3	0	3	2	1	9
15	10	15	10	16	16	18	15

Exercise 6.6: Average Weights of Children in Glasgow

Below are data collected in 1907 from school children in Glascow.
There is no information available on the number of children examined.
School districts were chosen to represent homogeneous ethnic background and divided into "poor" and "rich" districts. Is there a
difference between the "poor" and "rich" districts with respect to

the average weights of children? If so, can you model that difference as a function of age?

Age	Boys		Girls	
	Poor Districts	Rich Districts	Poor Districts	Rich Districts
6	41	43	40	42
7	44	47	43	46
8	48	51	46	49
9	52	56	51	54
10	57	61	55	59
11	62	66	59	64
12	66	71	65	71
13	72	77	72	79
14	76	83	77	89

Exercise 6.7: Cadmium Concentrations in Urine Samples

These data were taken from a calibration study involving five laboratories and the use of a new highly sensitive method for detecting cadmium in human fluids. Each urine sample was divided in five aliquots, so each lab tested all 13 samples. Is there any lab (or labs) that stands out as different from the rest? How "accurately" can this method measure cadmium concentration?

Cadmium concentrations in 13 urine samples as determined by five laboratories:

Sample Number	Estimated Cd μg/L				
	Lab 1	Lab 2	Lab 3	Lab 4	Lab 5
1	16.6	12.7	14.6	12	12.3
2	16.0	12.4	18.6	15	12.0
3	24.0	10.4	9.3	7	7.3
4	12.0	8.9	9.3	7	6.0
5	16.0	14.6	24.8	18	16.5
6	190.4	150.7	154.8	182	152.9
7	53.4	48.1	51.6	47	49.9
8	173.6	138.3	128.3	148	139.2
9	68.4	55.5	53.2	48	36.8
10	39.4	28.9	35.5	41	35.6
11	5.4	2.4	4.5	6	4.0
12	128.0	125.8	108.4	93	136.7
13	22.0	13.1	12.9	20	13.2

Exercise 6.8:   Comparative Occurrence of Letters in German and
                English

In 1936, J. B. Leishman published the German text with parallel
English translation of Rainer Maria Rilke's *Sonnets to Orpheus*
(Hogarth Press, London). Each page has an appropriate stanza of a
sonnet in German on the left and a very careful English translation
on the right. The number of "*i*'s" and "*t*'s" occurring in the first
50 words of each of the first 10 sonnets were counted and are recorded below. What can be said about the relative use of these two
letters in the two modern languages?

| | Counts of Letters | | | |
| | German | | English | |
Sonnet	"*i*"	"*t*"	"*i*"	"*t*"
I	20	9	14	14
II	20	6	10	14
III	23	13	11	17
IV	25	19	12	22
V	25	14	13	12
VI	23	12	15	13
VII	14	10	12	19
VIII	20	13	18	16
IX	23	19	12	18
X	23	19	12	18

Exercise 6.9:   Cloud Seeding Experiment

On each flight, two clouds were chosen as "good candidates" for
seeding. A prepared envelope was opened and the random number
inside dictated which would be seeded and which would be left
alone. The rain volume was estimated from each of the two clouds.
Sixteen such experiments were recorded. Does seeding do any good,
as indicated by these data? Does it do any harm (i.e., reduce the
rainfall)?

Rain volume from cloud seeding experiments:

Paired Trial	Seeded	Control
1	129.6	26.1
2	31.4	26.3
3	2745.6	87.0
4	489.1	95.0
5	430.0	372.4
6	302.8	0.0
7	119.0	17.3
8	4.1	24.4
9	92.4	11.5
10	17.5	321.2
11	200.7	68.5
12	274.7	81.2
13	283.7	47.3
14	7.7	28.6
15	1656.0	830.1
16	978.0	345.5

Milton Keynes UK
Ingram Content Group UK Ltd.
UKHW051953071024
449327UK00026B/2288